Multivariable Calculus

Kevin Woolsey

ii

Contents

Chapter 1

Vectors and Three Dimensional Space

1.1   Euclidean Space

For your entire math life, you have probably worked in the two dimensional xy plane. Now it's time to move up to the big leagues and add a *third* dimension. A third axis, the z axis, now stands perpendicular to the xy plane, with all three axes intersecting at the origin.

Notice that any point in three dimensional space requires three coordinates to describe it. In addition to a point's position on the xy plane, we must also give the point's height above or below this plane, which is the z coordinate.

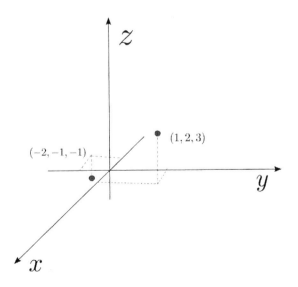

Figure 1.1: The standard way to draw three dimensional space, showing the orientation of the x, y, and z axes as well as some example points.

In addition to the xy plane, we will now also refer to the xz and yz planes, which are the planes where y is always 0 and x is always 0, respectively. Another term which will be used often is a *projection*. This can have multiple meanings depending on the context, but it is basically the shadow that one object casts onto another. For example, the projection of the point $(2, 4, 6)$ onto the xy plane would be $(2, 4, 0)$.

One more thing worth mentioning is the equivalent of quadrants in the xy plane. In three dimensions, the axes divide space into 8 *octants*, four above the xy plane and four below it. The first octant is the one in which all three coordinates of a point would be positive, i.e. the one directly facing you in the diagram above.

Sets of Real Numbers

A *set* is basically just a collection of objects called *elements*. You have probably seen the symbol $\mathbb{R}$ before, which denotes the set of all real numbers. We can also think of the set $\mathbb{R}$ as describing one dimensional space, the real number line. To describe any point in one dimensional space, you only need one number, and all single real numbers are elements of $\mathbb{R}$. So the elements of $\mathbb{R}$ can be interpreted as points in one dimensional space.

Going up one level, $\mathbb{R}^2$ denotes the set of all *pairs* of real numbers. Examples of elements in this set would be $(2, 5)$, $(-1, 7)$, or any other pair. So we see that elements of $\mathbb{R}^2$ can be interpreted as points in the xy plane.

Lastly, $\mathbb{R}^3$ describes three dimensional space, and this set consists of all *triples* of real numbers, e.g. $(2, 6, 3)$. We can keep going up to any number of dimensions. In general, $\mathbb{R}^n$ denotes the set of all n-tuples of real numbers. For example, an element of $\mathbb{R}^5$ is of the form $(2, 5, 3, 1, 9)$. Throughout the book, we will use this kind of notation often, especially when talking about functions.

Simple Shapes

To get a feel for three dimensional space, we look at some shapes represented by basic equations. When we talk about the visual representation of an equation, such as $z = 5$, we are looking for all possible points which satisfy the conditions set by the equation. In this case, any point with a z coordinate with 5 will satisfy the equation; the x and y coordinates don't matter. The collection of all points with a z coordinate of 5 form a horizontal *plane*. As another example, $x = 3$ represents a vertical plane consisting of all points whose x coordinate is 3.

For a slightly harder example, consider the equation $x^2 + y^2 = 1$. In two dimensional space, we know that this equation represents the unit circle. In three dimensional space, which points will satisfy the equation? Notice that z doesn't appear in the equation, so we only have restrictions on the x and y coordinates. The points on the unit circle in the xy plane obviously still satisfy the equation, and since the z coordinate doesn't matter, we essentially slide the unit circle up and down the z axis to form a cylinder.

The last shape we will look at is spheres, but first we need the three dimensional **_distance formula_**. The distance between two points (x_1, y_1, z_1) and (x_0, y_0, z_0) is defined as

$$\text{distance} = \sqrt{(x_1 - x_0)^2 + (y_1 - y_0)^2 + (z_1 - z_0)^2}$$

A sphere is defined as the set of all points which have the same distance, the radius, from the center. Therefore, if the center of the sphere is (x_0, y_0, z_0) and the radius is R, then all points on the sphere must satisfy the equation

$$(x - x_0)^2 + (y - y_0)^2 + (z - z_0)^2 = R^2$$

Example. Find the center and radius of the sphere $x^2 - 2x + y^2 + 6y + z^2 = 3$. If we can get it into the form above, the center and radius are immediately recognizable. To accomplish this, we complete the square.

$$(x^2 - 2x + 1) - 1 + (y^2 + 6y + 9) - 9 + z^2 = 3$$
$$(x - 1)^2 + (y + 3)^2 + z^2 = 13$$

So we see that the sphere has a center at $(1, -3, 0)$ and radius of $\sqrt{13}$.

Exercises

1. What shape does the equation $y = x$ describe in $\mathbb{R}^2$? What about in $\mathbb{R}^3$?

2. What shape does the equation $(x - 2)^2 + (y + 1)^2 = 9$ describe in $\mathbb{R}^3$?

3. What region (in $\mathbb{R}^3$) is described by the inequality $x^2 + y^2 \leq 4$? What about $x^2 + y^2 < 4$?

4. Find the distance between the points $(1, 2, 5)$ and $(3, 0, 4)$

5. Find an equation for the sphere centered at $(2, 3, 1)$ which just touches the xz plane at one point

6. Find the coordinates of a point in the first octant which is equidistant from all three coordinate axes

Answers

1. In two dimensional space, it describes the line passing through the origin with slope 1. In three dimensional space, it describes the vertical plane which stretches above and below this line, since z is not involved.

2. A cylinder of radius 3 obtained by sliding the circle centered at $(2, -1, 0)$ up and down

3. All points on or inside the cylinder of radius 2; all points inside the cylinder but not on the border

4. distance is 3

5. The sphere will have a radius of 3 because then it touches the xz plane at $(2, 0, 1)$, so the equation is $(x-2)^2 + (y-3)^2 + (z-1)^2 = 9$

6. Any point where all 3 coordinates are the same

1.2   Vectors

You have probably worked with vectors at least a little, but we will do a quick review of the basics in this section. To put it simply, a vector is an object which represents a *displacement* in space. The difference between a point and a vector is the difference between saying "I am at position $(2, 3)$" and "I moved 2 units to the right and 3 units up".

This latter statement can be represented mathematically by the vector $\mathbf{v} = \langle 2, 3 \rangle$, which is a two dimensional vector. The first number tells you how much to move in the x direction, the second number is the displacement in the y direction, and $\mathbf{v}$ is just the vector's name. The numbers 2 and 3 are called the *components* of the vector. In this book, we will denote vectors in bold, but note that many people draw an arrow on top to indicate a vector $\vec{v}$.

All vectors have a *magnitude* and a *direction*. The magnitude is the length of the vector; in other words, it tells you how far to move in the specified direction. To find the magnitude of a vector, take the square root of the sum of the squares of its components. For example, if we had a three dimensional vector $\mathbf{v} = \langle v_1, v_2, v_3 \rangle$, then its magnitude is given by

$$|\mathbf{v}| = \sqrt{v_1^2 + v_2^2 + v_3^2}$$

where the bars around the vector indicate magnitude. Note that while this is the same symbol as for absolute value, there can be no confusion because we never take the absolute value of a vector. Also, since a vector is completely characterized by its magnitude and direction, any two vectors with the same magnitude and direction are considered to be equal, regardless of their starting points. The only vector without a specific direction is the zero vector, which is $\mathbf{0} = \langle 0, 0, 0 \rangle$ in three dimensions.

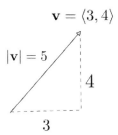

Vector Operations

Vectors have two important operations associated with them: vector addition and scalar multiplication. To add two vectors, we just add their corresponding components individually. For example,

$$\langle 2, 4 \rangle + \langle 3, 7 \rangle = \langle 5, 11 \rangle$$

This makes intuitive sense if we think about it in terms of displacements. If you first move 2 to the right and 4 up, then move 3 to the right and 7 up, in total you have moved 5 right and 11 up. To find the sum of two vectors geometrically, position the tail of the second vector at the tip of the first one. Their sum is then the vector which points from the tail of the first vector to the tip of the second.

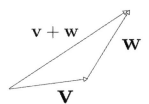

Vector addition has many properties of real number addition, for example the associative and commutative properties.

$$(\mathbf{x} + \mathbf{y}) + \mathbf{z} = \mathbf{x} + (\mathbf{y} + \mathbf{z})$$

$$\mathbf{x} + \mathbf{y} = \mathbf{y} + \mathbf{x}$$

These properties are implied by the way we defined vector addition (adding the components is just adding real numbers).

Scalar multiplication involves multiplying a vector by a real number, which is what a scalar is. To perform the multiplication, we simply multiply each component of the vector by the number. As an example,

$$3\langle 2, 6, 1 \rangle = \langle 6, 18, 3 \rangle$$

Multiplying a vector by a *positive* scalar only changes the length of the vector, not the direction. However, multiplying a vector by a negative number will make it point in the opposite direction as well as modify the length.

Now that we have both of these operations, we can also define the difference between two vectors $\mathbf{v} - \mathbf{w}$, which is really just multiplying the second vector by -1 then adding. Geometrically, the difference of two vectors is the vector which points from the tip of the second vector to the tip of the first vector, assuming that the vectors start from the same point. This can be seen from the identity

$$\mathbf{w} + (\mathbf{v} - \mathbf{w}) = \mathbf{v}$$

which tells us that the difference vector is the thing you add to the second one to get the first one.

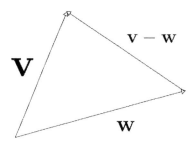

Unit Vectors

A **unit vector** is any vector with a magnitude of 1. Sometimes we want to turn a vector into a unit vector while still preserving its direction. To accomplish this, all we have to do is multiply the vector by the reciprocal of its magnitude.

$$\frac{\mathbf{v}}{|\mathbf{v}|}$$

An important class of unit vectors is the **standard basis vectors**. These vectors, denoted by $\mathbf{e}$ and a subscript, point in the direction of a coordinate axis. Therefore, in n dimensional space, there are n standard basis vectors. In three dimensions, they are

$$\mathbf{e}_1 = \langle 1, 0, 0 \rangle$$
$$\mathbf{e}_2 = \langle 0, 1, 0 \rangle$$
$$\mathbf{e}_3 = \langle 0, 0, 1 \rangle$$

Sometimes they are called **i**, **j**, and **k**, respectively. The significance of the standard basis vectors is that any vector can be written as a sum of the standard basis vectors. For example,

$$\langle 2, 7, 4 \rangle = 2\mathbf{e}_1 + 7\mathbf{e}_2 + 4\mathbf{e}_3$$

Points and Vectors

If we take any point, such as $(2, 4, 3)$, and the vector whose components correspond to the point's coordinates, $\langle 2, 4, 3 \rangle$, notice that the point will be at the tip of the vector when the vector starts at the origin. Therefore, any *point* can be represented by a *vector* that starts at the origin and has components corresponding to the point's coordinates. We say that the vector is the ***position vector*** of the point.

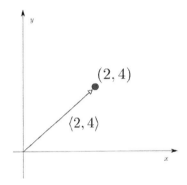

Because we can represent points by vectors, we will sometimes treat them as such if it is convenient. For example, if we had a point (a_1, a_2, a_3) we might instead write **a**. Just keep in mind that if it looks like we just replaced a point by a vector, we are actually just talking about the point's position vector.

To end the section, we revisit the distance formula. Now that we've looked at vectors, we can see that the distance formula can be derived from the magnitude of a vector. If we have two points $\mathbf{a} = \langle x_1, y_1, z_1 \rangle$ and $\mathbf{b} = \langle x_0, y_0, z_0 \rangle$ represented by their position vectors, then their difference $\mathbf{a} - \mathbf{b}$ is a vector which goes between the two points. Therefore, the magnitude of this vector should be the distance between the points.

$$|\mathbf{a} - \mathbf{b}| = \sqrt{(x_1 - x_0)^2 + (y_1 - y_0)^2 + (z_1 - z_0)^2}$$

Exercises

1. Consider the vectors in $\mathbb{R}^3$ $\mathbf{v} = \langle 1, 3, 2 \rangle$ and $\mathbf{w} = \langle 4, 0, 7 \rangle$. Perform the following operations

    (a) $2\mathbf{v} + 3\mathbf{w}$

    (b) $|\mathbf{v}| \, \mathbf{w}$

(c) $\mathbf{v} - 2\mathbf{w}$

2. Prove that, for any vector $\mathbf{v}$ and real number c,

$$|c\mathbf{v}| = |c|\,|\mathbf{v}|$$

3. Verify that the vector obtained by

$$\frac{1}{|\mathbf{v}|}\langle v_1, v_2, v_3\rangle$$

   is indeed a unit vector

   (a) Find two unit vectors parallel to $\mathbf{v} = \langle 2, -, 1, 3\rangle$

4. Write out the standard basis vectors in $\mathbb{R}^4$

Answers

1. (a) $\langle 14, 6, 25\rangle$

   (b) $\langle 4\sqrt{14}, 0, 7\sqrt{14}\rangle$

   (c) $\langle -7, 3, -12\rangle$

3. (a) $\pm\frac{1}{\sqrt{14}}\langle 2, -1, 3\rangle$

4. $\mathbf{e}_1 = \langle 1, 0, 0, 0\rangle, \mathbf{e}_2 = \langle 0, 1, 0, 0\rangle, \mathbf{e}_3 = \langle 0, 0, 1, 0\rangle, \mathbf{e}_4 = \langle 0, 0, 0, 1\rangle$

1.3   The Dot Product

Notions such as distance and angle are intuitive to us in two or three dimensions, but things start to become unclear in higher dimensions. What does distance mean in five dimensions? What does it mean for two vectors to be perpendicular in nine dimensions? Considering these questions, we would like a new way to define these concepts. The tool that we will use to describe these geometric properties of space is the **dot product**, also called the inner product or scalar product.

The dot product is an operation we perform on two vectors to yield a *scalar*. Given two vectors $\mathbf{a}$ and $\mathbf{b}$, their dot product is defined as the sum of the products of their corresponding components. For example, if they are three dimensional vectors,

$$\mathbf{a} \cdot \mathbf{b} = a_1 b_1 + a_2 b_2 + a_3 b_3$$

Example. If $\mathbf{a} = \langle 2, 4, 3\rangle$ and $\mathbf{b} = \langle 1, 5, 0\rangle$,

$$\mathbf{a} \cdot \mathbf{b} = (2)(1) + (4)(5) + (3)(0) = 22$$

The first thing we note is that the dot product of a vector with itself is equal to the square of its magnitude.

$$\mathbf{a} \cdot \mathbf{a} = a_1 a_1 + a_2 a_2 + a_3 a_3$$
$$= a_1^2 + a_2^2 + a_3^2$$
$$= |\mathbf{a}|^2$$

In fact, the magnitude of a vector is *defined* as the square root of the dot product with itself. This gives us a new way to define distance, since in the last section we saw that we can interpret distance as the magnitude of a vector. Next we want to define the notion of angle, in particular the angle between two vectors. To do this, we need a new identity.

Theorem. If θ is the angle between $\mathbf{a}$ and $\mathbf{b}$,

$$\mathbf{a} \cdot \mathbf{b} = |\mathbf{a}| \, |\mathbf{b}| \cos \theta$$

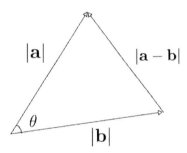

Proof. Consider the triangle made up of the vectors $\mathbf{a}$, $\mathbf{b}$, and $\mathbf{a} - \mathbf{b}$. The lengths of the triangle's sides are the magnitudes of these vectors. Applying the law of cosines to this triangle,

$$|\mathbf{a} - \mathbf{b}|^2 = |\mathbf{a}|^2 + |\mathbf{b}|^2 - 2 |\mathbf{a}| \, |\mathbf{b}| \cos \theta$$
$$(\mathbf{a} - \mathbf{b}) \cdot (\mathbf{a} - \mathbf{b}) = |\mathbf{a}|^2 + |\mathbf{b}|^2 - 2 |\mathbf{a}| \, |\mathbf{b}| \cos \theta$$
$$(\mathbf{a} \cdot \mathbf{a}) - 2(\mathbf{a} \cdot \mathbf{b}) + (\mathbf{b} \cdot \mathbf{b}) = |\mathbf{a}|^2 + |\mathbf{b}|^2 - 2 |\mathbf{a}| \, |\mathbf{b}| \cos \theta$$
$$|\mathbf{a}|^2 - 2(\mathbf{a} \cdot \mathbf{b}) + |\mathbf{b}|^2 = |\mathbf{a}|^2 + |\mathbf{b}|^2 - 2 |\mathbf{a}| \, |\mathbf{b}| \cos \theta$$
$$-2(\mathbf{a} \cdot \mathbf{b}) = -2 |\mathbf{a}| \, |\mathbf{b}| \cos \theta$$
$$\mathbf{a} \cdot \mathbf{b} = |\mathbf{a}| \, |\mathbf{b}| \cos \theta$$

$\square$

In the proof we used the fact that the distributive property holds for the dot product, which we will talk about later. This formula gives us an alternate way to calculate the dot product and also a way to *define* the angle between any two vectors:

$$\theta = \arccos \left(\frac{\mathbf{a} \cdot \mathbf{b}}{|\mathbf{a}| \, |\mathbf{b}|} \right)$$

Notice that if the angle between two vectors is $\frac{\pi}{2}$, then their dot product is zero. This motivates us to define two vectors to be ***orthogonal*** (same as perpendicular) if their dot product is zero. As a consequence of this definition, the zero vector $\mathbf{0}$ is orthogonal to every vector.

Properties

We will now look at various properties of the dot product. The proofs all use three dimensional vectors, but the properties hold for any vectors.

1. $\mathbf{a} \cdot \mathbf{b} = \mathbf{b} \cdot \mathbf{a}$ (commutative)

$$\mathbf{a} \cdot \mathbf{b} = a_1 b_1 + a_2 b_2 + a_3 b_3$$
$$= b_1 a_1 + b_2 a_2 + b_3 a_3$$
$$= \mathbf{b} \cdot \mathbf{a}$$

2. $\mathbf{a} \cdot (\mathbf{b} + \mathbf{c}) = \mathbf{a} \cdot \mathbf{b} + \mathbf{a} \cdot \mathbf{c}$ (distributive)

$$\mathbf{a} \cdot (\mathbf{b} + \mathbf{c}) = \langle a_1, a_2, a_3 \rangle \cdot \langle b_1 + c_1, b_2 + c_2, b_3 + c_3 \rangle$$
$$= a_1(b_1 + c_1) + a_1(b_2 + c_2) + a_3(b_3 + c_3)$$
$$= a_1 b_1 + a_2 b_2 + a_3 b_3 + a_1 c_1 + a_2 c_2 + a_3 c_3$$
$$= \mathbf{a} \cdot \mathbf{b} + \mathbf{a} \cdot \mathbf{c}$$

3. $c(\mathbf{a} \cdot \mathbf{b}) = (c\mathbf{a}) \cdot \mathbf{b} = \mathbf{a} \cdot (c\mathbf{b})$ where c is a real number

$$c(\mathbf{a} \cdot \mathbf{b}) = ca_1 b_1 + ca_2 b_2 + ca_3 b_3$$
$$= (ca_1)b_1 + (ca_2)b_2 + (ca_3)b_3 = (c\mathbf{a}) \cdot \mathbf{b}$$
$$= a_1(cb_1) + a_2(cb_2) + a_3(cb_3) = \mathbf{a} \cdot (c\mathbf{b})$$

Projections

Often times we want to find out how much one vector points in the direction of a second vector. In other words, given vectors $\mathbf{a}$ and $\mathbf{b}$, we want to know what the magnitude of $\mathbf{a}$ in the direction of $\mathbf{b}$ is. This quantity is called the ***scalar projection*** of $\mathbf{a}$ onto $\mathbf{b}$.

As you can see in the picture, the length of $\mathbf{a}$ in the direction of $\mathbf{b}$ seems to be the quantity $|\mathbf{a}| \cos \theta$, where θ is the angle between the vectors. We can use the dot product formula to get rid of the θ in this expression.

$$\mathbf{a} \cdot \mathbf{b} = |\mathbf{a}|\,|\mathbf{b}| \cos \theta$$

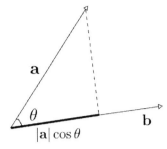

Figure 1.2: The scalar projection of **a** onto **b**.

$$|\mathbf{a}| \cos\theta = \frac{\mathbf{a} \cdot \mathbf{b}}{|\mathbf{b}|}$$

This last expression is what we will use to calculate the scalar projection. The notation for the scalar projection of **a** onto **b** is sometimes written as

$$\text{comp}_\mathbf{b}(\mathbf{a}) = \frac{\mathbf{a} \cdot \mathbf{b}}{|\mathbf{b}|}$$

We can also define a ***vector projection***, which is basically the scalar projection turned into a vector. That is, the vector projection of **a** onto **b** is a vector which has a magnitude equal to the scalar projection and points in the direction of **b**. To calculate this, we first find the unit vector in the direction of **b**, then multiply this by the scalar projection of **a** onto **b**. The notation is sometimes written as

$$\text{proj}_\mathbf{b}(\mathbf{a}) = \left(\frac{\mathbf{a} \cdot \mathbf{b}}{|\mathbf{b}|}\right)\frac{\mathbf{b}}{|\mathbf{b}|}$$

Example. Given $\mathbf{a} = \langle 2, 4, 1 \rangle$ and $\mathbf{b} = \langle 1, 5, 3 \rangle$, find the scalar and vector projections of **a** onto **b**. First, we calculate all the necessary components.

$$\mathbf{a} \cdot \mathbf{b} = 2 + 20 + 3 = 25$$

$$|\mathbf{b}| = \sqrt{1 + 25 + 9} = \sqrt{35}$$

Directly using the formula, the scalar projection is

$$\text{comp}_\mathbf{b}(\mathbf{a}) = \frac{\mathbf{a} \cdot \mathbf{b}}{|\mathbf{b}|} = \frac{25}{\sqrt{35}}$$

The unit vector in the direction of **b** is

$$\frac{\mathbf{b}}{|\mathbf{b}|} = \frac{1}{\sqrt{35}}\langle 1, 5, 3 \rangle$$

Therefore, the vector projection is

$$\text{proj}_\mathbf{b}(\mathbf{a}) = \left(\frac{25}{\sqrt{35}}\right)\frac{1}{\sqrt{35}}\langle 1, 5, 3 \rangle = \frac{5}{7}\langle 1, 5, 3 \rangle = \langle \frac{5}{7}, \frac{25}{7}, \frac{15}{7} \rangle$$

Exercises

1. Consider the vectors $\mathbf{a} = \langle -1, 3, 5 \rangle$ and $\mathbf{b} = \langle 2, -2, 3 \rangle$

   (a) Compute $\mathbf{a} \cdot \mathbf{b}$

   (b) Calculate, in degrees, the angle between the two vectors

   (c) Find the scalar projection of $\mathbf{a}$ onto $\mathbf{b}$

   (d) Find the vector projection of $\mathbf{a}$ onto $\mathbf{b}$

   (e) Find a vector orthogonal to $\mathbf{b}$ such that $\mathbf{a}$ can be written as the sum of this vector and the vector projection

2. Show that if the dot product of $\mathbf{a}$ and $\mathbf{b}$ is positive, then the angle between the vectors is acute; if the dot product is negative, the angle is obtuse.

3. Prove the Cauchy Schwarz inequality $|\mathbf{a} \cdot \mathbf{b}| \leq |\mathbf{a}|\,|\mathbf{b}|$

4. Use the Cauchy Schwarz inequality and dot product properties to prove the triangle inequality $|\mathbf{a}| + |\mathbf{b}| \leq |\mathbf{a}| + |\mathbf{b}|$ (Hint: start with $|\mathbf{a+b}|^2$)

   (a) Why is this called the triangle inequality?

5. The direction angles α, β, and γ are defined as the angles a vector makes with the x, y, and z axes, respectively. Find expressions for these angles given a vector $\mathbf{v} = \langle v_1, v_2, v_3 \rangle$

Answers

1. (a) 7

   (b) $\theta \approx 73.3°$

   (c) $\frac{7}{\sqrt{17}}$

   (d) $\frac{7}{17}\langle 2, -2, 3 \rangle$

   (e) subtract the vector projection from $\mathbf{a}$

2. Arccos returns an angle between 0 and $\frac{\pi}{2}$ for positive numbers and an angle between $\frac{\pi}{2}$ and π for negative numbers, and $\theta = \arccos\left(\frac{\mathbf{a} \cdot \mathbf{b}}{|\mathbf{a}||\mathbf{b}|}\right)$

3. Use the formula $\mathbf{a} \cdot \mathbf{b} = |\mathbf{a}|\,|\mathbf{b}|\cos\theta$. Cosine is always between -1 and 1, so $-|\mathbf{a}|\,|\mathbf{b}| \leq \mathbf{a} \cdot \mathbf{b} \leq |\mathbf{a}|\,|\mathbf{b}|$ from which the inequality follows

4.

$$|\mathbf{a}| + |\mathbf{b}|^2 = (\mathbf{a} + \mathbf{b}) \cdot (\mathbf{a} + \mathbf{b})$$
$$= |\mathbf{a}|^2 + 2(\mathbf{a} \cdot \mathbf{b}) + |\mathbf{b}|^2$$
$$\leq |\mathbf{a}|^2 + 2\,|\mathbf{a}|\,|\mathbf{b}| + |\mathbf{b}|^2 \qquad \text{(Cauchy Schwarz)}$$
$$= (|\mathbf{a}| + |\mathbf{b}|)^2$$

The inequality follows once you square root both sides of our final expression

(a) Imagine a triangle made up of vectors $\mathbf{a}$, $\mathbf{b}$, and $\mathbf{a} + \mathbf{b}$. The inequality states that one side of the triangle cannot be greater than the sum of the other two sides.

5. $\alpha = \arccos(\frac{v_1}{|\mathbf{v}|})$, $\beta = \arccos(\frac{v_2}{|\mathbf{v}|})$, $\gamma = \arccos(\frac{v_3}{|\mathbf{v}|})$. Apply the angle formula with the standard basis vectors.

1.4 The Cross Product

In the last section, we saw that the dot product of two vectors produces a scalar, and we can use it to describe many geometrical properties of vectors. In this section, we will look at another operation on vectors which produces a new vector instead of a scalar. Before that, however, we need to look at some properties of determinants.

Determinants

A determinant can be thought of as a function of the rows (or columns) of a *square* matrix which returns a real number. In our discussion of the properties of determinants, you can always interchange the words "row" and "column". For a 2×2 matrix, the determinant is

$$\begin{vmatrix} a_1 & a_2 \\ b_1 & b_2 \end{vmatrix} = a_1 b_2 - a_2 b_1$$

For a 3×3 matrix, the determinant is

$$\begin{vmatrix} a_1 & a_2 & a_3 \\ b_1 & b_2 & b_3 \\ c_1 & c_2 & c_3 \end{vmatrix} = a_1 \begin{vmatrix} b_2 & b_3 \\ c_2 & c_3 \end{vmatrix} - a_2 \begin{vmatrix} b_1 & b_3 \\ c_1 & c_3 \end{vmatrix} + a_3 \begin{vmatrix} b_1 & b_2 \\ c_1 & c_2 \end{vmatrix}$$

There are a few things you should note here. First, don't forget the negative sign in front of the second term. Also, the little 2×2 matrices appearing next to the a's in the 3×3 formula are called *submatrices*, and they are obtained by crossing out the row and column which the a is in. For example, in the first term, you get that submatrix by crossing out the first row and first column because that's where a_1 is.

Besides how to compute 2×2 and 3×3 determinants, all you need to know about them is a couple properties. I will not give the proofs here because the general proofs would require some extra concepts that we will not use again. The proofs for 2×2 and 3×3 determinants, however, can be done through direct computation, which is no fun.

1. If two rows are the same or multiples of each other, then the determinant is automatically zero.

$$\begin{vmatrix} 2 & 3 & 2 \\ 1 & 7 & 6 \\ 2 & 3 & 2 \end{vmatrix} = 0$$

2. Switching two rows has the same effect as multiplying the value of the determinant by -1

$$\begin{vmatrix} 1 & 2 & 3 \\ 5 & 1 & 2 \\ 7 & 6 & 1 \end{vmatrix} = - \begin{vmatrix} 5 & 1 & 2 \\ 1 & 2 & 3 \\ 7 & 6 & 1 \end{vmatrix}$$

3. Multiplying a row by a constant c has the same effect as multiplying the value of the determinant by c

$$\begin{vmatrix} 5 & 6 \\ 2 & 4 \end{vmatrix} = 2 \begin{vmatrix} 5 & 6 \\ 1 & 2 \end{vmatrix}$$

The Cross Product

Now we can move on to the main topic, the cross product. The cross product takes two vectors and produces a new vector, and unlike the dot product this operation is *only defined for three dimensional vectors*. For two vectors in $\mathbb{R}^3$ **a** and **b**, their dot product is defined as

$$\mathbf{a} \times \mathbf{b} = \begin{vmatrix} a_2 & a_3 \\ b_2 & b_3 \end{vmatrix} \mathbf{e}_1 - \begin{vmatrix} a_1 & a_3 \\ b_1 & b_3 \end{vmatrix} \mathbf{e}_2 + \begin{vmatrix} a_1 & a_2 \\ b_1 & b_2 \end{vmatrix} \mathbf{e}_3$$

Noticing the similarity of this definition to the formula of a 3×3 determinant, we often write

$$\mathbf{a} \times \mathbf{b} = \begin{vmatrix} \mathbf{e}_1 & \mathbf{e}_2 & \mathbf{e}_3 \\ a_1 & a_2 & a_3 \\ b_1 & b_2 & b_3 \end{vmatrix}$$

However, this is technically not a true determinant because only numbers are allowed to be in a determinant, not vectors.

Example. Compute the cross product of $\mathbf{a} = \langle 2, 1, 3 \rangle$ and $\mathbf{b} = \langle 1, 0, 4 \rangle$.

$$\mathbf{a} \times \mathbf{b} = \begin{vmatrix} 1 & 3 \\ 0 & 4 \end{vmatrix} \mathbf{e}_1 - \begin{vmatrix} 2 & 3 \\ 1 & 4 \end{vmatrix} \mathbf{e}_2 + \begin{vmatrix} 2 & 1 \\ 1 & 0 \end{vmatrix} \mathbf{e}_3$$

$$= (4 - 0)\mathbf{e}_1 - (8 - 3)\mathbf{e}_2 + (0 - 1)\mathbf{e}_3$$

$$= \langle 4, -5, -1 \rangle$$

Properties

The cross product is useful because the new vector it creates is always *orthogonal* to the two original vectors. To prove this, we need to show that the dot product between $\mathbf{a} \times \mathbf{b}$ and both $\mathbf{a}$ and $\mathbf{b}$ equals 0, since two vectors are orthogonal if their dot product is zero.

$$\mathbf{a} \cdot (\mathbf{a} \times \mathbf{b}) = a_1 \begin{vmatrix} a_2 & a_3 \\ b_2 & b_3 \end{vmatrix} - a_2 \begin{vmatrix} a_1 & a_3 \\ b_1 & b_3 \end{vmatrix} + a_3 \begin{vmatrix} a_1 & a_2 \\ b_1 & b_2 \end{vmatrix}$$

$$= \begin{vmatrix} a_1 & a_2 & a_3 \\ a_1 & a_2 & a_3 \\ b_1 & b_2 & b_3 \end{vmatrix}$$

But the determinant is zero if it contains two identical rows, so the dot product of $\mathbf{a}$ and $\mathbf{a} \times \mathbf{b}$ is zero and they are orthogonal. The same argument can be repeated for the dot product with $\mathbf{b}$. This proof also shows us that if we have the dot product of a vector and a cross product $\mathbf{a} \cdot (\mathbf{b} \times \mathbf{c})$, we can compute it by replacing the first row of the cross product definition with the first vector $\mathbf{a}$.

$$\mathbf{a} \cdot (\mathbf{b} \times \mathbf{c}) = \begin{vmatrix} a_1 & a_2 & a_3 \\ b_1 & b_2 & b_3 \\ c_1 & c_2 & c_3 \end{vmatrix}$$

As a side note, notice that there are two possible directions which are orthogonal to both $\mathbf{a}$ and $\mathbf{b}$. There is a way to determine which way the cross product will point if you cannot do the actual calculations. According to the *right hand rule*, to find the direction of the cross product you point your index finger in the direction of the first vector, your middle finger towards the second, and then stick your thumb straight up. Your thumb will then be pointing along the cross product. You can practice with this picture:

We now list some important properties of the cross product.

1. $\mathbf{a} \times \mathbf{b} = -(\mathbf{b} \times \mathbf{a})$ (NOT commutative)

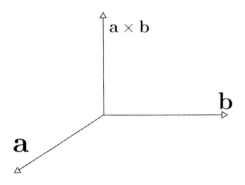

This property follows from the fact that switching two rows of a determinant multiplies the whole thing by -1.

$$\mathbf{a} \times \mathbf{b} = \begin{vmatrix} \mathbf{e}_1 & \mathbf{e}_2 & \mathbf{e}_3 \\ a_1 & a_2 & a_3 \\ b_1 & b_2 & b_3 \end{vmatrix} = - \begin{vmatrix} \mathbf{e}_1 & \mathbf{e}_2 & \mathbf{e}_3 \\ b_1 & b_2 & b_3 \\ a_1 & a_2 & a_3 \end{vmatrix} = -(\mathbf{b} \times \mathbf{a})$$

2. $\mathbf{a} \times (\mathbf{b} + \mathbf{c}) = (\mathbf{a} \times \mathbf{b}) + (\mathbf{a} \times \mathbf{b})$ (distributive)

We didn't talk about this property earlier because it is mostly irrelevant for our purposes, but if one row in a determinant is the sum of two rows, then you can split it up like this:

$$\begin{vmatrix} \mathbf{e}_1 & \mathbf{e}_2 & \mathbf{e}_3 \\ a_1 & a_2 & a_3 \\ b_1 + c_1 & b_2 + c_2 & b_3 + c_3 \end{vmatrix} = \begin{vmatrix} \mathbf{e}_1 & \mathbf{e}_2 & \mathbf{e}_3 \\ a_1 & a_2 & a_3 \\ b_1 & b_2 & b_3 \end{vmatrix} + \begin{vmatrix} \mathbf{e}_1 & \mathbf{e}_2 & \mathbf{e}_3 \\ a_1 & a_2 & a_3 \\ c_1 & c_2 & c_3 \end{vmatrix}$$

3. $c(\mathbf{a} \times \mathbf{b}) = (c\mathbf{a}) \times \mathbf{b} = \mathbf{a} \times (c\mathbf{b})$ where c is a real number (same property that dot product has)

This works because multiplying one row of a determinant by a constant c is the same thing as multiplying the whole determinant by c.

4. $\mathbf{a} \cdot (\mathbf{b} \times \mathbf{c}) = (\mathbf{a} \times \mathbf{b}) \cdot \mathbf{c}$

This type of expression involving the dot product of a vector and a cross product is called a **scalar triple product**. In words, this property says that the positions of the dot and the cross in a scalar triple product can be switched with no effect on the value. This is because switching rows of a determinant two times has no effect (since you multiply by -1 for each switch).

$$\begin{vmatrix} a_1 & a_2 & a_3 \\ b_1 & b_2 & b_3 \\ c_1 & c_2 & c_3 \end{vmatrix} = \begin{vmatrix} c_1 & c_2 & c_3 \\ a_1 & a_2 & a_3 \\ b_1 & b_2 & b_3 \end{vmatrix}$$

There is also a convenient formula to compute the *magnitude* of the cross product, similar to the alternate formula for the dot product involving cosine.

$$|\mathbf{a} \times \mathbf{b}| = |\mathbf{a}|\,|\mathbf{b}|\sin\theta$$

where θ again is the angle between the two vectors. The proof of this formula is very tedious, so we will just give an outline of the steps and you can go through the computations if you have nothing better to do.

$$
\begin{aligned}
|\mathbf{a}|^2\,|\mathbf{b}|^2 \sin^2\theta &= |\mathbf{a}|^2\,|\mathbf{b}|^2 - |\mathbf{a}|^2\,|\mathbf{b}|^2\cos^2\theta \\
&= |\mathbf{a}|^2\,|\mathbf{b}|^2 - (\mathbf{a}\cdot\mathbf{b})^2 \quad \text{(remember dot product formula)} \\
&= (a_1^2 + a_2^2 + a_3^2)(b_1^2 + b_2^2 + b_3^2) - (a_1 b_1 + a_2 b_2 + a_3 b_3)^2 \\
&= (a_2 b_3 - a_3 b_2)^2 + (a_1 b_3 - a_3 b_1)^2 + (a_1 b_2 - a_2 b_1)^2 \\
&= |\mathbf{a}\times\mathbf{b}|^2
\end{aligned}
$$

Taking the square root of both sides then gives you the identity. For the dot product, if the value was 0, then that meant the two vectors were orthogonal. Since the cross product magnitude involves $\sin\theta$, we see that the cross product is 0 if the angle is 0 or π. In other words, if the cross product of two vectors equals 0, then the vectors are *parallel*.

This formula also has a geometrical meaning: the magnitude of the cross product of two vectors is the *area of the parallelogram* determined by the two vectors. This follows from some trigonometry and the fact that the area of a parallelogram is the length of its base times its height.

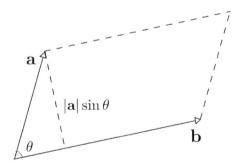

Lastly, we note that while the cross product is only defined for three dimensional vectors, sometimes we can cheat and do it for two dimensional vectors by extending them to three dimensions. We basically just add a 0 in the third component and think of them as lying completely in the xy plane in three dimensional space.

Example. If we have $\mathbf{a} = \langle 4,5 \rangle$ and $\mathbf{b} = \langle 2,3 \rangle$, then we can extend them to three dimensions as $\mathbf{a} = \langle 4,5,0 \rangle$ and $\mathbf{b} = \langle 2,3,0 \rangle$. Then

$$\mathbf{a}\times\mathbf{b} = \langle 0,0,2 \rangle$$

The cross product of two dimensional vectors will always only point in the z direction, which makes sense because it has to be orthogonal to the xy plane containing both original vectors.

Exercises

1. Calculate the determinant

    (a)
    $$\begin{vmatrix} 2 & 1 & 3 \\ 7 & 0 & 4 \\ 6 & 2 & 1 \end{vmatrix}$$

    (b)
    $$\begin{vmatrix} 1 & 5 & 6 & 2 \\ 6 & 9 & 6 & 2 \\ 3 & 1 & 4 & 7 \\ 1 & 5 & 6 & 2 \end{vmatrix}$$

2. Prove the following determinant properties for a 2×2 determinant

    (a) If two rows are the same or multiples of each other, the determinant is zero

    (b) Switching two rows has the same effect as multiplying the determinant by -1

    (c) Multiplying a row by a constant c has the same effect as multiplying the determinant by c

3. Consider the vectors $\mathbf{a} = \langle 7, 2, 1 \rangle$ and $\mathbf{b} = \langle 3, 5, 2 \rangle$

    (a) Compute $\mathbf{a} \times \mathbf{b}$

    (b) Find the area of the parallelogram determined by $\mathbf{a}$ and $\mathbf{b}$

    (c) Find the angle between the two vectors. Use both the cross product and dot product formulas, and check that they give the same answer.

4. For two vectors, their dot product is equal to the magnitude of their cross product. What is the angle between the vectors?

5. A parallelepiped is a three dimensional shape whose faces are all parallelograms. Consider the parallelepiped determined by vectors $\mathbf{a}$, $\mathbf{b}$, and $\mathbf{c}$, where $\mathbf{a}$ and $\mathbf{b}$ form the base of the solid and $\mathbf{c}$ points above them, forming a diagonal edge. The volume of the parallelepiped is the area of its base times its height.

    (a) Find an expression for the area of the base

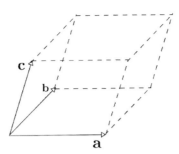

(b) Find an expression for the height (Hint: it involves finding the scalar projection of **c** onto something)

(c) Show that the volume of the parallelepiped is given by the *absolute value* of the scalar triple product $(\mathbf{a} \times \mathbf{b}) \cdot \mathbf{c}$

Answers

1. (a) 43

   (b) 0 because the first and last rows are equal

2. (a)
$$\begin{vmatrix} a_1 & a_2 \\ ca_1 & ca_2 \end{vmatrix} = (a_1)(ca_2) - (a_2)(ca_1) = 0$$

   (b)
$$\begin{vmatrix} b_1 & b_2 \\ a_1 & a_2 \end{vmatrix} = b_1 a_2 - b_2 a_1 = -(a_1 b_2 - a_2 b_1)$$

   (c)
$$\begin{vmatrix} ca_1 & ca_2 \\ b_1 & b_2 \end{vmatrix} = ca_1 b_2 - ca_2 b_1 = c(a_1 b_2 - a_2 b_1)$$

3. (a) $\langle -1, -11, 29 \rangle$

   (b) $3\sqrt{107}$

   (c) $43.2°$

4. $\frac{\pi}{4}$; equate the dot product and cross product formulas to find that $\cos\theta = \sin\theta$

5. (a) $|\mathbf{a} \times \mathbf{b}|$

   (b) The cross product points perpendicular to the base, so the height can be found by the scalar projection of **c** onto the cross product $\frac{(\mathbf{a} \times \mathbf{b}) \cdot \mathbf{c}}{|\mathbf{a} \times \mathbf{b}|}$

   (c) Multiply the answers to parts (a) and (b), and the magnitude of the cross product cancels out. The absolute value is needed in order for the formula to work for any configuration of the vectors. For example, if the cross product happened to point in the opposite direction as **c**, the scalar projection would be negative.

1.5   Equations of Lines and Planes

Now that we have worked with vectors and their two types of products, we can start using them to describe different objects in three dimensional space. Some of the simplest objects are straight lines and planes, which can both be characterized by a particular vector.

Lines

To describe a line in two or three dimensional space, all we need is a point on the line, which we will refer to as **a** (remember points can be represented by their position vectors), and *any* vector **v** which is parallel to the line, called the ***direction vector***. To see why, look at the following picture:

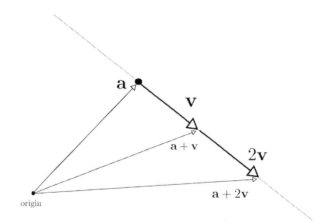

When we add the vectors **a** and **v**, we get to another point on the line, again represented by its position vector. If we multiply **v** by a constant before adding them, we get a different point on the line. We can get to *any* point on the line by adding some constant multiple of **v** to **a**. If we call this constant multiple t, then all points on the line can be obtained from the following equation by plugging in the right value of t.

$$L(t) = \mathbf{a} + t\mathbf{v}$$

So when we plug a value of t into this ***vector equation*** of the line, it returns the position vector of a point on the line which corresponds to that t value. Now suppose that the initial point is $\mathbf{a} = \langle x_0, y_0, z_0 \rangle$ and the direction vector is $\mathbf{v} = \langle a, b, c \rangle$. Then the vector equation becomes

$$L(t) = \langle x_0 + at, y_0 + bt, z_0 + ct \rangle$$

We can interpret the individual components of the vector equation as giving us the x, y, and z coordinates of the point at a certain value of t. We can take the components as separate equations and call them the ***parametric equations*** for the line.

$$x(t) = x_0 + at$$

$$y(t) = y_0 + bt$$

$$z(t) = z_0 + ct$$

Example. Find an equation for the line which passes through the points $(2, 0, 3)$ and $(5, 2, 1)$. All we need is one point on the line, which could be either of the two given, and any vector parallel to the line. For the direction vector, we can choose the vector which points between these two points. The vector that goes from the first point to the second point is

$$\mathbf{v} = \langle 3, 2, -2 \rangle$$

Therefore, the vector equation of this line is

$$L(t) = \langle 2, 0, 3 \rangle + t\langle 3, 2, -2 \rangle$$

and the parametric equations are

$$x(t) = 2 + 3t$$

$$y(t) = 2t$$

$$z(t) = 3 - 2t$$

In two dimensional space, two lines either intersect or are parallel (meaning their direction vectors are parallel). In three dimensional space, however, two lines can be ***skew***: not intersecting yet also not parallel.

Example. Line one has an initial point $\mathbf{a}_1 = \langle 2, 3, 5 \rangle$ and direction vector $\mathbf{v}_1 = \langle -5, 0, 0 \rangle$. Line two has an initial point $\mathbf{a}_2 = \langle 2, -3, 0 \rangle$ and direction vector $\mathbf{v}_2 = \langle 1, -2, 4 \rangle$. Show that these lines are skew.

First of all, you can see that the lines are not parallel because their direction vectors are not constant multiples of each other and are therefore not parallel. We then need to show that there is no point which lies on both of these lines. We use s instead of t for line two because we don't want to confuse them; they both just represent the independent variable of the vector equations.

$$L_1(t) = \langle 2 - 5t, 3, 5 \rangle$$

$$L_2(s) = \langle 2 + s, -3 - 2s, 4s \rangle$$

If the lines intersect, then there exist values of t and s which yield the exact same point when we plug them into the equations of lines one and two. From the third components, we see that the only possible time when the z coordinates of points on the lines can match is when $s = \frac{5}{4}$. However, looking at the second components, the only time when the y coordinates can match is when $s = -3$. Since s cannot be both of these values at once, we conclude that the lines do not intersect and are therefore skew, since they are not parallel either.

Planes

To describe a plane in three dimensional space, all we need is a point on the plane and a ***normal vector***, that is, a vector which is perpendicular to the plane. Again, we will represent this initial point on the plane with its position vector $\mathbf{a}$. We want an equation which every point on the plane must fulfill, so let's bring in a random point on the plane $\mathbf{x} = \langle x, y, z \rangle$. The vector pointing from the initial point to any point on the plane $\mathbf{x} - \mathbf{a}$ will lie on the plane and therefore be orthogonal to our normal vector $\mathbf{n}$. If two vectors are orthogonal, then their dot product is zero, so every point on the plane $\mathbf{x}$ (or the position vector of every point) must satisfy this equation:

$$(\mathbf{x} - \mathbf{a}) \cdot \mathbf{n} = 0$$

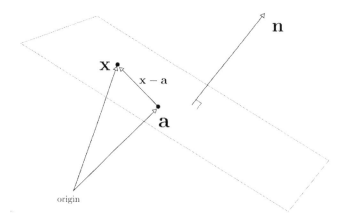

If we write the initial point as $\mathbf{a} = \langle x_0, y_0, z_0 \rangle$ and the normal vector as $\mathbf{n} = \langle a, b, c \rangle$, then the equation becomes

$$\langle x - x_0, y - y_0, z - z_0 \rangle \cdot \langle a, b, c \rangle = 0$$

$$a(x - x_0) + b(y - y_0) + c(z - z_0) = 0$$

Note a couple things about this equation. First, anytime you are given an equation of a plane, to find a normal vector all you have to do is take the coefficients of x, y, and z. Second, there is a difference in the way we described planes and lines. For a line, we came up with a vector equation which gave us a point on the line every time we plugged in a different value of t. For planes, we don't have any plugging in business; we have an equation which all points on the plane must satisfy. Lastly, two planes in three dimensional space must either intersect or be parallel, which means that their normal vectors are parallel.

Example. Find an equation of the plane which passes through the three points $P(2, 3, 1)$, $Q(4, 0, 3)$, and $R(1, 5, 2)$.

For the initial point, we can choose any of these points because they are all on the plane. The problem is finding a normal vector. Notice that the any of the vectors between these three points will all lie on the plane, so any normal vector must be orthogonal to them. Therefore, if we find two vectors which are both on the plane, we can take their cross product, which will be orthogonal to both original vectors and therefore the plane as well. We take the vectors pointing between P and Q and P and R, but you can do other combinations as well.

$$PQ = \langle 2, -3, 2 \rangle$$

$$PR = \langle -1, 2, 1 \rangle$$

$$\mathbf{n} = PQ \times PR = \langle -7, -4, 1 \rangle$$

Choosing P as our initial point, an equation of the plane is

$$-7(x - 2) - 4(y - 3) + (z - 1) = 0$$

Exercises

1. Given the plane $3x - 7y + 2z = 10$, find two normal vectors which point in opposite directions

   (a) Find an equation for a line which is perpendicular to this plane

2. Find an equation for the plane which contains the point $P(1, 2, 1)$ and is orthogonal to the vector $\langle 3, 4, 1 \rangle$

   (a) Given the point $Q(2, 3, 2)$, show that this point is not on the plane, and find a vector from any point on the plane to this new point

(b) Find the shortest distance between the plane and the point Q (Hint: it involves scalar projections)

3. Consider plane one $2x + 3y - z = 6$ and plane two $4x + 6y - 2z = 10$. Show that the planes are parallel.

    (a) Find the shortest distance between the planes

4. Consider the planes $3x + 2y + z = 5$ and $5x - 3y + 2z = 3$. Show that the planes are not parallel, so they must intersect

    (a) Find the angle that the planes intersect at

    (b) Find an equation for the line of intersection between the planes

5. Consider the points $P(2, 1, 4)$, $Q(1, 2, 3)$, and $R(5, 1, 1)$

    (a) Find an equation of the plane which passes through these three points

    (b) Find the distance between this plane and the point $S(3, 6, 2)$

    (c) Find the angle of intersection between this plane and the plane $7(x - 2) + 2(y - 1) + 4(z + 3) = 0$

    (d) Find an equation for the line of intersection between these two planes

    (e) Does this line intersect the line $L(s) = \langle \frac{-8}{3}, \frac{22}{3}, 6 \rangle + s\langle 2, -1, -5 \rangle$? If so, find the point of intersection

6. Consider the lines $L_1(t) = \langle 0, 0, 0 \rangle + t\langle 1, 1, 1 \rangle$ and $L_2(s) = \langle -1, 0, 0 \rangle + s\langle 1, 2, 3 \rangle$

    (a) Show that the lines are skew

    (b) Find the shortest distance between the lines (Hint: we need to find a vector which is always orthogonal to both lines so we can project something on it)

    (c) Find the distance between L_2 and the point $(5, 3, 6)$

Answers

1. $\langle 3, -7, 2 \rangle$ and $\langle -3, 7, -2 \rangle$, or any scalar multiple of these two

    (a) Find any point on the plane, for example $(0, 0, 5)$, and take any normal vector to the plane as a direction vector for the line. Example answer: $L(t) = \langle 0, 0, 5 \rangle + t\langle 3, -7, 2 \rangle$

2. $3(x - 1) + 4(y - 2) + (z - 1) = 0$

(a) The point does not satisfy the equation of the plane. An example of such a vector would be $PQ = \langle 1, 1, 1 \rangle$.

(b) $\frac{8}{\sqrt{26}}$; Find the scalar projection of the vector from part (a) onto the normal vector. The vector in part (a) represents some displacement from the plane to Q, so the scalar projection onto the normal vector tells you how much of this displacement is perpendicular to the plane, and the shortest possible displacement from the plane to the point will be perpendicular.

3. Their normal vectors are scalar multiples of each other and therefore parallel $\mathbf{n}_1 = \langle 2, 3, -1 \rangle$, $\mathbf{n}_2 = 2\mathbf{n}_1$

   (a) $\frac{1}{\sqrt{14}}$;The method is essentially the same as in exercise 2(b). Find any vector which goes from a point in one plane to a point in the other plane, and find the scalar projection of this vector onto one of the normal vectors. The distance is the absolute value of this quantity.

4. Their normal vectors are not parallel (not scalar multiples of each other)

   (a) 61.5°; This angle is the same thing as the angle between their normal vectors (draw a picture)

   (b) The line is running perpendicular to both normal vectors of the planes, so taking the cross products of the normal vectors provides us with a direction vector for the line. Then you just need a point which is on this line, i.e. on both planes. Example answer: $L(t) = \langle 0, 1, 3 \rangle + t\langle 7, -1, -19 \rangle$

5. (a) Example answer: $(x - 2) + 2(y - 1) + (z - 4) = 0$

   (b) $\frac{9}{\sqrt{6}}$

   (c) 42.5°

   (d) To get a direction vector, cross the two normal vectors. To get a point on both planes, it might help to set up a system of linear equations like so

   $$x + 2y + z = 8$$

   $$7x + 2y + 4z = 4$$

   which can be solved by picking a random value of z then solving for x and y by elimination/substitution. Example answer: $L(t) = \langle -\frac{2}{3}, \frac{13}{3}, 0 \rangle + t\langle 2, 1, -4 \rangle$

   (e) yes, $(\frac{4}{3}, \frac{16}{3}, -4)$; set up a system of equations in terms of t and s by equating the corresponding components of the lines' vector equations, then solve

6. (a) They are not parallel because their direction vectors are not parallel and they also have no point of intersection

(b) $\frac{1}{\sqrt{6}}$; Take the cross product of the direction vectors to obtain a vector orthogonal to both lines. Find a vector from a point on one line to a point on the other, e.g. $\langle -1, 0, 0 \rangle$ and find the (absolute value of) scalar projection of this onto that cross product.

(c) $\sqrt{\frac{117}{7}}$; First, find a vector from the line to the point, e.g. $\langle 6, 3, 6 \rangle$. Find the vector projection of this vector onto the direction vector $\frac{15}{7}\langle 1, 2, 3 \rangle$. Subtract this from the first displacement vector you had $\langle \frac{27}{7}, -\frac{9}{7}, -\frac{3}{7} \rangle$. This represents the shortest perpendicular displacement from the line to the point, so its magnitude is the distance.

1.6   Quadric Surfaces

In this section, we will examine various types of surfaces in three dimensional space and the equations which describe them. Visualizing these shapes can be tough, but we will look at different techniques to determine surfaces from equations so you don't have to simply memorize which equation goes with what.

In general, a **_quadric surface_** is any surface which is represented by an equation involving x, y, and z to at most the second power, e.g. $2x^2 + 3xy + z = 5$ or $x + y^2 - z = 2$. However, we will only look at special types quadric surfaces which are given names.

Cross Sections

The most common tool used to analyze quadric surface equations is cross sections, also called traces. A cross section is one slice of a surface; in other words, the intersection of the surface with a particular plane. To illustrate this technique we will use the equation

$$z = x^2 + y^2$$

Let's see what the cross sections perpendicular to the z axis look like. The cross section of the surface at the plane $z = 1$ is the curve we would get if we took a slice of the surface at $z = 1$. To obtain this, we simply set $z = 1$ in the equation

$$1 = x^2 + y^2$$

So this cross section is a circle of radius 1 in the xy plane (at a height of 1). We can do this for an arbitrary cross section at $z = k$, where k is just a constant.

$$k = x^2 + y^2$$

We see that all of the cross sections perpendicular to the z axis are circles, and as we climb to higher z values, the circles get larger and larger.

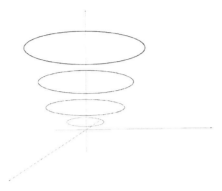

Figure 1.3: Cross sections perpendicular to the z axis of the surface $z = x^2 + y^2$

To complete the picture, we can find the cross sections perpendicular to the x and y axes. For slices perpendicular to the x axis at a value $x = k$,

$$z = y^2 + k^2$$

So these cross sections are upward parabolas in the yz plane (at $x = k$) whose vertices get higher and higher as x increases. If we did cross sections perpendicular to the y axis, we would get the same parabola shapes. The actual quadric surface looks like a bowl:

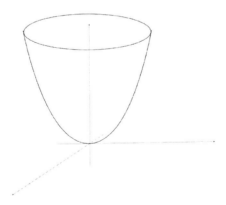

This quadric surface is called a ***paraboloid***, or more specifically a circular paraboloid. We can also have elliptic paraboloids which are the same but have ellipses for cross sections perpendicular to the z axis.

Example. Use cross sections to analyze the shape of the surface given by $z = x^2 - y^2$.

First, we do cross sections perpendicular to the z axis. When the surface intersects with the plane $z = k$, it looks like

$$k = x^2 - y^2$$

$$1 = \frac{x^2}{k} - \frac{y^2}{k}$$

These cross sections are hyperbolas in the xy plane at a height of k. For positive values of z, the hyperbolas are of the form $1 = \frac{x^2}{k} - \frac{y^2}{k}$ so they open up left and right. For negative values of z, they take the form $1 = \frac{y^2}{k} - \frac{x^2}{k}$ and therefore open up and down.

The cross sections perpendicular to the x axis are downward parabolas in the yz plane, and the cross sections perpendicular to the y axis are upward parabolas in the xz plane.

$$z = k^2 - y^2 \qquad \text{and} \qquad z = x^2 - k^2$$

This is the hardest quadric surface to visualize, so you probably still have no idea what it looks like after doing these cross sections.

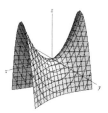

This saddle or pringle shaped surface is called a ***hyperbolic paraboloid***.

Rotations

I'm not a big fan of cross sections, so we're going to move on to the next technique: rotational symmetry. All of the quadric surfaces we will study have rotational symmetry except the hyperbolic paraboloid, so this is a very effective tool. To illustrate this technique, we will use the surface represented by

$$x^2 + y^2 - z^2 = 1$$

The key idea here is that this equation can be written entirely in terms of z and $\sqrt{x^2 + y^2}$. We will call the latter quantity r_z because geometrically it gives the distance of a point from the z axis.

$$r_z^2 - z^2 = 1$$

The fact that we can rewrite the equation to only involve r_z and z implies that the surface has rotational symmetry around the z axis. Why? If a collection of points on the surface have the same z coordinate, the equation implies that they all have the same distance from the z axis.

Since the surface has rotational symmetry about the z axis, we can obtain the surface by rotating some curve around the z axis. Our job now is to find a suitable curve to rotate. There are multiple ways to do this, but a convenient way is to look at the cross section of the surface with the plane $y = 0$. In other words, look at the curve in the half of the xz plane where x is positive, then rotate that around the z axis. To find this curve, we just set $y = 0$ in the equation

$$x^2 - z^2 = 1$$

We are only interested in positive x values, so this curve is half of a hyperbola which opens left and right. We rotate this around the z axis and the resulting shape is our quadric surface. This one is called a **hyperboloid of one sheet**.

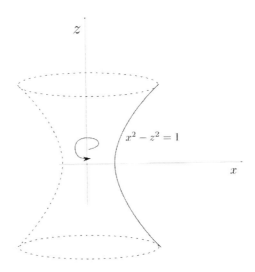

Example. Describe the surface given by $z^2 - x^2 - y^2 = 1$.

We see that it has rotational symmetry about the z axis because we can rewrite it in terms of z and the distance from the z axis $z^2 - r_z^2 = 1$. Therefore, all we need to do is find a curve to rotate around the z axis. As before, we will look for this curve in the half of the xz plane where x is positive. Setting $y = 0$, we obtain the curve

$$z^2 - x^2 = 1$$

You can see that this curve is similar to the example we just did, but the positions of the z and x are switched. This tells us that the curve in the xz plane is a hyperbola which opens up and down. Rotating this curve arounnd the z axis, we get two small bowl shapes. This quadric surface is called a **hyperboloid of two sheets**.

This rotational symmetry technique doesn't only work for rotations around the z axis. If we can write a function in terms of y and the distance from the y axis $\sqrt{x^2 + z^2}$, then the surface must have rotational symmetry about the y axis. Similarly, if we can express the function using only x and the distance from the x axis $\sqrt{y^2 + z^2}$, then the surface has rotational symmetry about the x axis. The key, then, is looking for the right curve to rotate about the axis in question.

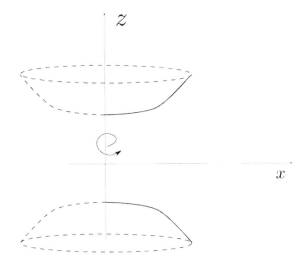

Example. Describe the surface given by $x^2 + z^2 = y^2$.

Immediately, you can see that this surface will have rotational symmetry about the y axis because it can be written in terms of y and the distance from the y axis $r_y^2 = y^2$. There are many ways to look for a curve to rotate about the y axis, but a convenient choice seems to be the half of the xy plane where x is positive. In the xy plane, $z = 0$, so we plug this into the equation and find our curve.

$$x^2 = y^2$$

$$y = \pm x$$

So our curve actually consists of two straight lines, one with slope 1 and the other with slope -1. Rotating this around the y axis gets us a **cone** centered on the y axis.

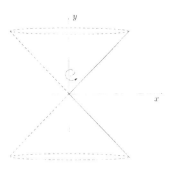

Transformations

The last thing we will look at is something you are actually already familiar with. If you are given an equation such as

$$(x - 2)^2 + y^2 = 1$$

you immediately recognize it as a circle of radius 1 whose center is $(2, 0)$. In other words, changing the x to $x - 2$ in the equation of the unit circle shifted the circle to the right 2 units. So making some change to one of the variables in the equation changes the shape in the opposite way. To use an example in three dimensions,

$$z = (x+1)^2 + (y-3)^2$$

This equation represents a circular paraboloid, whose basic equation is $z = x^2 + y^2$, which has been shifted 1 unit in the negative x direction and 3 units in the positive y direction. So even though we can't directly apply rotational symmetry to the above equation, we can recognize it as a transformation of the equation $z = x^2 + y^2$, which we can use rotational symmetry with.

Another common transformation is a stretch or a squish, which are achieved by multiplying a variable by a coefficient. For example, the parabola $y = 2x$ is the parabola $y = x$ squished by a factor of 2 in the x direction (it will be steeper). So multiplying by a variable by a number squishes the shape in the direction of that variable. If the number is a fraction, it stretches the shape in that direction. Looking at things in this light, we can interpret an ellipse as simply a circle which has been stretched in the x and y directions.

$$\frac{x^2}{9} + \frac{y^2}{16} = 1$$

In this ellipse, we started with the unit circle and then multiplied x and y by $\frac{1}{3}$ and $\frac{1}{4}$, respectively.

$$x^2 + y^2 = 1$$

$$\left(\frac{x}{3}\right)^2 + \left(\frac{y}{4}\right)^2 = 1$$

So we just have a circle which has been stretched by a factor of 3 in the x direction and a factor of 4 in the y direction.

Example. Describe the surface given by

$$\frac{x^2}{9} + \frac{y^2}{16} + (z-1)^2 = 1$$

We recognize this as the unit sphere $x^2 + y^2 + z^2 = 1$ with some transformations applied. First, we have a stretch by a factor of 3 and 4 in the x and y directions. Next, we have a shift up one unit in the positive z direction.

This deformed sphere is called an ***ellipsoid***.

Exercises

1. Describe the cross sections perpendicular to the x, y, and z axes of the hyperbolic paraboloid $z = y^2 - x^2$

2. Use rotational symmetry to describe the following surfaces

    (a) $z^2 = x^2 + y^2$

    (b) $y = x^2 + z^2$

    (c) $x^2 - y^2 - z^2 = 1$

    (d) $x^2 + z^2 - y^2 = 1$

3. Describe the transformations which have been applied to the "base shape"

    (a) $x^2 + (2y - 1)^2 + z^2 = 1$

    (b) $z = 2x^2 + (y + 2)^2$

Answers

1. For $x = k$, upward parabolas in the yz plane; for $y = k$, downward parabolas in the xz plane; for $z = k$, hyperbolas in xy plane: up/down for positive k and left/right for negative k

2.   (a) A cone centered around the z axis

    (b) A circular paraboloid centered around the y axis

    (c) Hyperboloid of two sheets centered around the x axis

    (d) Hyperboloid of one sheet centered aorund the y axis

3.   (a) The unit sphere was first shifted 1 unit in the positive y direction, then squished by a factor of 2 in the y direction. Notice that if you applied the transformations in the opposite order, you would get

$$(2y)^2$$

$$(2(y - 1))^2 = (2y - 2)^2$$

    (b) The circular paraboloid was squished by a factor of $\sqrt{2}$ in the x direction and shifted 2 units in the negative y direction

1.7   Cylindrical and Spherical Coordinates

In some situations, rectangular coordinates (the ones we've been using) will be a little too square for our needs. Take, for example, the unit circle in the xy plane given by

$$x^2 + y^2 = 1$$

This doesn't seem too complicated, but look how much simpler the equation becomes in polar coordinates:

$$r = 1$$

This happens often in three dimensions when we want to describe certain curvy surfaces.

Cylindrical Coordinates

The first new three dimensional coordinate system we examine is ***cylindrical coordinates***, which is literally just an extension of polar coordinates.

To describe a point, we use the three coordinates (r, θ, z). r and θ are the same thing as in polar coordinates; we use them to describe the point's projection onto the xy plane. The z coordinate is the same as in rectangular coordinates, denoting the height of the point above or below the xy plane. Conversion between rectangular and cylindrical coordinates is quite simple because z is the same, and for x and y we can use the same polar coordinate conversions that you already know.

$$x = r \cos \theta$$

$$y = r \sin \theta$$

$$x^2 + y^2 = r^2$$

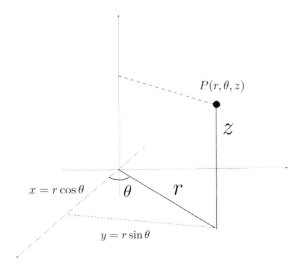

When working with polar coordinates before, you might have let r be negative and θ take on any value so that the coordinates of any point could be written in an infinite number of ways (by adding 2π to θ). While there are no "offical" restrictions on these parameters, in this book we will say r must be positive and θ only ranges between 0 and 2π. There is pretty much no situation in which it would be more convenient to lift these restrictions.

Example. Convert the point $(1, \sqrt{3}, 5)$ from rectangular coordinates to cylindrical coordinates.

We first find that $r = 2$ because $r^2 = x^2 + y^2 = 1 + 3 = 4$. Now we find θ by looking for a value which satisfies both equations

$$1 = 2\cos\theta$$

$$\sqrt{3} = 2\sin\theta$$

The only possible value seems to be $\theta = \pi/3$, so the point in cylindrical coordinates is $(2, \frac{\pi}{3}, 5)$

The reason it's called cylindrical coordinates is because the equation for a cylinder (centered along the z axis) becomes very simple. A cylinder can be described as the collection of points which all have the same distance from the z axis, which is exactly what the parameter r in cylindrical coordinates describes. For example, the equation for a cylinder of radius 4 becomes

$$x^2 + y^2 = 16$$

$$r^2 = 16$$

$$r = 4$$

Spherical Coordinates

The other system we look at is ***spherical coordinates***. We describe a point using the three parameters (ρ, θ, ϕ). ρ is the distance between the origin and the point, θ is the same thing as in cylindrical coordinates, and ϕ is the angle between the positive z axis and the point's position vector. Some people use different symbols, so note that θ is also called the azimuthal angle and ϕ is also called the zenith angle.

Some useful conversion equations, which you can derive by staring at the picture hard enough, are

$$x = \rho\sin\phi\cos\theta$$

$$y = \rho\sin\phi\sin\theta$$

$$z = \rho\cos\phi$$

$$x^2 + y^2 + z^2 = \rho^2$$

As with cylindrical coordinates, we will place some restrictions on the range of the parameters. We say that ρ must be positive, θ can range from 0 to 2π, and ϕ can range from 0 (straight up the z axis) to π (straight down the z axis). These restrictions still allow us to describe any point using spherical coordinates and will prevent unnecessary confusion later on.

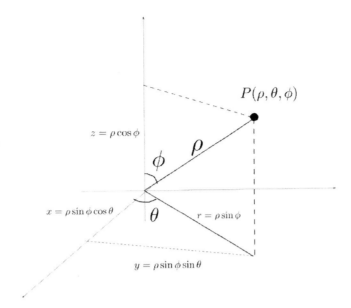

Example. Convert the point $(\frac{5\sqrt{2}}{4}, \frac{5\sqrt{2}}{4}, \frac{5\sqrt{3}}{2})$ from rectangular to spherical coordinates.

The easiest thing to do is to find $\rho = 5$ because

$$x^2 + y^2 + z^2 = \frac{50}{16} + \frac{50}{16} + \frac{75}{4} = \frac{400}{16} = 25$$

The next easiest thing to find is ϕ from the z coordinate because the formula for the z coordinate only involves two parameters.

$$\frac{5\sqrt{3}}{2} = 5\cos\phi$$

Therefore $\phi = \pi/6$. Lastly, we need to find the value of θ which satisfies both equations

$$\frac{5\sqrt{2}}{4} = 5(\frac{1}{2})\cos\theta$$

$$\frac{5\sqrt{2}}{4} = 5(\frac{1}{2})\sin\theta$$

The only value that works here is $\pi/4$, so the point in spherical coordinates is $(5, \frac{\pi}{4}, \frac{\pi}{6})$.

Lastly, we note that this system is called spherical coordinates because the equation of a sphere (centered around the origin) is very simple. A sphere is the collection of points that have the same distance from the center, so if the center is the origin then the parameter ρ describes this distance. For example, the equation for the sphere centered at the origin of radius 6 becomes

$$\rho = 6$$

Exercises

1. Convert the point $(-1, -1, 7)$ from rectangular to cylindrical coordinates

2. Convert the point $(-1, \sqrt{3}, -2\sqrt{3})$ from rectangular to spherical coordinates

3. Describe the surface whose equation is given in cylindrical or spherical coordinates

   (a) $z = r$

   (b) $\phi = \frac{\pi}{4}$

   (c) $r^2 + z^2 = 4$

4. Describe the region in three dimensional space generated as the given parameters range between the given values

   (a) $0 \le r \le 1,\ 0 \le \theta \le \pi,\ 0 \le z \le 1$

   (b) $0 \le \rho \le 4,\ 0 \le \theta \le 2\pi,\ 0 \le \phi \le \frac{\pi}{2}$

Answers

1. $(\sqrt{2}, \frac{5\pi}{4}, 7)$

2. $(4, \frac{2\pi}{3}, \frac{5\pi}{6})$

3. (a) A cone; whether it consists of two or one parts depends on if you let r be negative. If you restrict r to positive values, then it is only the half of the cone which is above the xy plane.

   (b) A cone; all the points are tilted an angle of $\frac{\pi}{4}$ away from the positive z axis; you could also convert this to cylindrical coordinates by using the fact $\phi = \arctan(r/z)$, from which you would derive $z = r$

   (c) Sphere of radius 2

4. (a) Half of a cylinder of radius 1 between the horizontal planes $z = 0$ and $z = 1$

   (b) The upper half of a sphere of radius 4

Chapter 2

Functions

2.1 Multivariable Functions

Intuitively, a function is just an object which takes an input, does some stuff to it, then gives you an output. Up until now, you have probably only worked with single variable functions, where the input is a single real number and the output is also a single real number. We often denote the input number by x and the output by y, so that $y = f(x)$.

The inputs for a particular function come from a set called the function's ***domain***, and the function assigns each element of the domain to a *unique* element in the ***range***. For a single variable function, both the domain and range are just $\mathbb{R}$ since the inputs and outputs are single real numbers. We can say that the function $y = f(x)$ maps elements of $\mathbb{R}$ to elements of $\mathbb{R}$, or $f : \mathbb{R} \mapsto \mathbb{R}$.

I emphasized unique because a function cannot assign a single point in the domain to multiple points in the range. For example, if $f(2) = 3$, then $f(2)$ cannot also equal 5. We can, however, have multiple *different* inputs mapping to the same output. So $f(5)$ could be equal to 3 also. Now we look at how to extend the idea of a function to allow for multiple inputs and multiple outputs.

Real Valued Functions

A class of functions which we will deal with often in multivariable calculus is ***real valued functions***. As the name suggests, real valued functions are functions which output a single real number, regardless of the type of input.

An example could be a two variable function $f : \mathbb{R}^2 \mapsto \mathbb{R}$. Notice that the domain of this function is $\mathbb{R}^2$; it takes a *pair* of real numbers as an input, and assigns that pair to a single real number as output. For example, a function like this could take $(4, 3)$ and output 1.

Often times when we have a two variable function we denote the pair of numbers that it takes as input by x and y. For a two variable real valued function, we often denote the output number by z, so that we can write $z = f(x, y)$.

Example. Evaluate $f(1, 2)$ if $f(x, y) = x^4 y^2 + x^2 y^4$

$$f(1, 2) = (1)^4 (2)^2 + (1)^2 (2)^4$$
$$= 4 + 8 = 12$$

We can have a function with any number of input variables. A function $f : \mathbb{R}^3 \mapsto \mathbb{R}$ takes a *triple* of real numbers as input and assigns them to a single real number. In this case we often denote the three independent variables by x, y, and z, so that the function is $f(x, y, z)$. You can see that if the dimension of our domain gets any higher, we will run out of letters to call our independent variables! For convenience, we often write the input of a multivariable function as a vector. So instead of writing $f(x, y, z)$ we might write $f(\mathbf{x})$. Another option you see a lot is assigning subscripts to the variables like so: $f(x_1, x_2, x_3, x_4)$, where $f : \mathbb{R}^4 \mapsto \mathbb{R}$.

Domains

Note that while we may write $f : \mathbb{R}^2 \mapsto \mathbb{R}$ for a two variable real valued function, usually the domain of the function will not be the entirety of $\mathbb{R}^2$ and the range will not be the entirety of $\mathbb{R}$. The function may be undefined at the point $(2, 5)$, or maybe the function only outputs positive numbers. Since all points in the domain are elements of $\mathbb{R}^2$, but it may not contain *all* pairs of real numbers, we say the domain is a *subset* of $\mathbb{R}^2$. Similarly, the range is a subset of $\mathbb{R}$. Keep that in mind, but we will still use the notation mainly to indicate the dimensions of the domain and range of a function.

For a function of two variables, where the domain is a subset of $\mathbb{R}^2$, each element or point in the domain can be visualized as a point on the xy plane, and the entire domain can be visualized as a region on the xy plane.

Example. Describe the domain of the real valued function $f(x, y) = \ln(y - x)$. Since the natural log is only defined for positive values, we obtain the restriction $y - x > 0$ or $y > x$. This implies that the domain is the region of the xy plane which is above, but not on, the line $y = x$.

For a function of three variables, the domain is a subset of $\mathbb{R}^3$, which can be visualized as a region of three dimensional space.

Example. Describe the domain of the real valued function $f(x, y, z) = \sqrt{4 - x^2 - y^2 - z^2}$. Since the square root cannot be negative, we obtain the restriction $x^2 + y^2 + z^2 \leq 4$. Therefore, the domain is all points in three dimensional space inside or on the sphere of radius

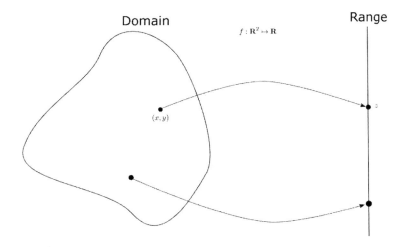

Figure 2.1: A real valued function of two variables. Points in the domain are a pair of numbers in a two dimensional region. The function assigns them to points in the range, which is one dimensional.

2 centered at the origin. You could also say the *closed ball* of radius 2 centered at the origin.

Vector Valued Functions

Lastly, we will extend our functions to include multiple output numbers. Any function whose range has a dimension higher than 1 is called a ***vector valued function***, or simply vector function, because we often think of its output as a vector. For example, the single variable function $f : \mathbb{R} \mapsto \mathbb{R}^2$

$$f(t) = \langle t, t^2 \rangle$$

takes in a single number t and spits out a *pair* of numbers (or a two dimensional vector).

Each component of the output vector is its own real valued functions, and together the components are called the ***coordinate functions***. In the above function, t and t^2 are the coordinate functions.

You actually already saw vector valued functions earlier in this book. The vector equation of a line is an example of a vector function; it took a number t and gave us the position vector of a point on the line.

Example. Evaluate $f(1,2)$ for the function $f : \mathbb{R}^2 \mapsto \mathbb{R}^3$ given by

$$f(x, y) = \langle x^2 y, 2x + y, y^3 \rangle$$

This function takes a pair of real numbers in $\mathbb{R}^2$ and assigns them to a triple of real numbers in $\mathbb{R}^3$. The output vector at $(1, 2)$ is

$$f(1, 2) = \langle 2, 4, 8 \rangle$$

As a side note, a vector function is considered to be undefined at a point in the domain if at least one of the coordinate functions is undefined there. In other words, an input value can only be in the domain of the overall vector function if it is also in the domain of all the coordinate functions.

Exercises

1. Evaluate the function $f : \mathbb{R}^4 \mapsto \mathbb{R}^2$ given by $f(x_1, x_2, x_3, x_4) = \langle 2x_2^2 - x_4, 5x_3 + 2x_1 \rangle$ at $(2, 4, 3, 1)$

2. Write an example of a function $f : \mathbb{R}^3 \mapsto \mathbb{R}^5$

3. Describe the domain of the real valued function

   (a) $f(x, y) = e^{\sqrt{y - x^2}}$

   (b) $f(x, y, z) = \ln(4 - x^2 - y^2 - z^2)$

   (c) $f(\mathbf{x}) = x_1^2 x_3 - 5x_4 + \sqrt{1 - x_1^2 - x_2^2 - x_3^2 - x_4^2}$

4. Find the domain of the vector valued function

$$f(t) = \langle \frac{1}{\sqrt{36 - t^2}}, t^3 - e^t, \ln|t| \rangle$$

5. If we have two functions $f : \mathbb{R} \mapsto \mathbb{R}^3$ and $g : \mathbb{R}^2 \mapsto \mathbb{R}$, is the composition of the functions $f \circ g$ possible? How about $g \circ f$?

Answers

1. $\langle 31, 19 \rangle$

2. Function takes a three dimensional vector as input and outputs a five dimensional vector; Example answer: $f(x, y, z) = \langle xy, 5z, 4x^2 z, y - z, 6y \rangle$

3. (a) $y \geq x^2$; all points on or above the parabola

   (b) $x^2 + y^2 + z^2 < 4$; all points within, but not on, the sphere of radius 2

   (c) $x_1^2 + x_2^2 + x_3^2 + x_4^2 \leq 1$; all points within or on the hypersphere in four dimensional space with radius 1

4. $-6 < t < 6, \quad t \neq 0$

5. $f \circ g$ is possible because the output of g and input of f match dimensions. $g \circ f$ is not possible because f outputs a triple of numbers which would not fit into g, who takes a pair of numbers

2.2 Visual Representations of Functions

In this section we will look at different ways to visually represent a function. For single variable real valued functions, we almost always use graphs. While this works fine for one or two variable functions, we will see that graphing a three or higher variable function or any vector valued function can be tricky and overall a waste of time. For these types of functions, other types of visuals will often be more useful.

Graphs

While graphs in general can refer to any sort of visual representation of stuff, in this section a "graph" will refer to the graph of a function.

To construct a graph for a particular function, we plot the points whose coordinates include both the inputs and outputs of the function. For example, take the function $f(x) = x^2$, whose graph you know to be a parabola in the xy plane. To find points to put on our graph, we take an input value, such as $x = 5$, and find its corresponding output value, $f(5) = 25$. We then plot the point $(5, 25)$ on the xy plane. Repeating this process for many input values creates a parabola in two dimensional space. Note that while the domain and range of the function are both one dimensional, the graph ends up being an object embedded in two dimensional space.

For a two variable real valued function $z = f(x, y)$, the graph consists of all points $(x, y, f(x, y))$. In other words, we take an input (x, y), find its corresponding output $f(x, y)$, then plot the point whose coordinates account for both the input and the output. The graph of this type of function then lives in three dimensional space, despite the domain being two dimensional and the range being one dimensional.

Since each point on the graph of a function needs to include all input variables and all output variables, the graph will live in a dimension equal to the sum of the dimensions of the domain and range. So you can see how this can get inconvenient for higher dimensional functions. If we had a three variable real valued function $f : \mathbb{R}^3 \mapsto \mathbb{R}$, its graph would live in four dimensional space, which we humans have a hard time visualizing.

Images

Recall from your previous classes that a parabola $y = x^2$ can be represented by the two parametric equations

$$x(t) = t \quad \text{and} \quad y(t) = t^2$$

Instead of writing out the two separate equations, we can combine them into a vector valued function $f : \mathbb{R} \mapsto \mathbb{R}^2$ given by

$$f(t) = \langle t, t^2 \rangle$$

When we plug in a value of t, this function spits out a pair of numbers (x, y) which can be interpreted as a point on the parabola. The *graph* of this function, however, is clearly not a parabola in the xy plane because according to our discussion of graphs above, the graph of this function should live in three dimensional space. So what exactly do we mean when we say that this function represents the parabola $y = x^2$?

Notice the wording used above: the function spits out a pair of numbers which represent a point on the parabola. So to get the parabola, we are just plotting the *outputs* of the function, not the inputs. When you look at a curve represented by a vector function such as this one, it is usually not clear which value of t in the domain produced a particular point on the curve.

The set of all outputs of a function is called its ***image***; we are therefore plotting all points in the image of the function $f(t) = \langle t, t^2 \rangle$ to produce the parabola. Since the image of a function is a collection of the points in the range of a function, the image will live in the same dimensional space as the range. The function we've been talking about maps $\mathbb{R}$ to $\mathbb{R}^2$, so its image, the parabola, lives in two dimensional space. If we had a vector function $f : \mathbb{R} \mapsto \mathbb{R}^3$, then its image would be a curve in three dimensional space.

Level Sets

The last type of visual representation we will talk about here is the ***level set***. A level set of a function is a collection of points *in the domain* which all correspond to the same output value. Therefore, we can have an infinite number of different level sets for any given function, each level set corresponding to a single output.

For example, let's say we have a function $T(x, y)$ which gives us the temperature at a point (x, y) on a piece of paper. I want to know all points on the paper where the temperature is 5, so I can set $T(x, y) = 5$. All points (x, y) on the paper which satisfy this equation will have a temperature of 5. If we plot this collection of points, we obtain a level set, which will look like a curve in the xy plane, of the function corresponding to the output 5. Many times we draw multiple level sets of a function together in the same picture.

Notice that to construct a level set we plot points in the *domain*, as opposed to an image where we plot points in the *range*. As a result, one advantage of drawing level sets is that they live in the same dimensional space as the domain of the function.

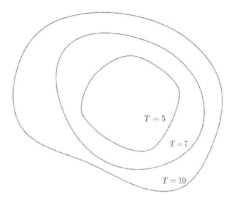

Figure 2.2: Three level sets of our temperature function $T(x, y)$. As you walk along any single curve, the temperature will stay constant.

Another example of a level set that you are familiar with is a circle. An entire circle cannot be the graph of a function because it fails the vertical line test. We can, however, interpret it as one level set of the function $f(x, y) = x^2 + y^2$. For example, the circle of radius 2 is really all points in the domain of the function which correspond to an output of 4, i.e. they satisfy $f(x, y) = 4$.

Example. Describe the level sets of the function $f(x, y, z) = x^2 + y^2 + z^2$.

This function has a three dimensional domain, so the level sets will live in three dimensional space. To determine what these level sets actually look like, we set the function equal to an arbitrary constant.

$$x^2 + y^2 + z^2 = k$$

We see that the level set which corresponds to an output value k is the sphere of radius $\sqrt{k}$ in three dimensional space. If we plotted multiple level sets together, they would be a bunch of concentric spheres centered around the origin.

To end the section, we note that any graph of a function can always be represented as both the image of some function and as a level set of some function. Say we have the graph of a two variable real valued function $z = f(x, y)$.

The graph, which consists of all points $(x, y, f(x, y))$, can be viewed as the *image* of the function $F : \mathbb{R}^2 \mapsto \mathbb{R}^3$ given by $F(x, y) = \langle x, y, f(x, y) \rangle$. It can also be viewed as the *level set* of the function $F : \mathbb{R}^3 \mapsto \mathbb{R}$ given by $F(x, y, z) = z - f(x, y)$ corresponding to the output value 0.

Exercises

1. How do we perform the vertical line test for the graph of a two variable real valued function $f(x, y)$?

2. Consider the function $f(x, y, z) = \langle f_1, f_2, f_3, f_4, f_5 \rangle$.

    (a) Write the function using the notation $f : \mathbb{R}^n \mapsto \mathbb{R}^m$

    (b) What dimension will the graph of the function live in?

    (c) The graph of the function itself will be a what dimensional object? For example, the graph of $y = x^2$ is a one dimensional curve living in two dimensional space.

    (d) What dimension will the image be in? What about a level set?

3. For each function, find a new function whose image matches the graph of the given function, and find another function which has a level set matching the graph of the function

    (a) $y = x^3 + 5$

    (b) $z = x^2 + y^2$

4. Describe the level sets of the given function

    (a) $f(x, y) = y - 2x$ for any constant k

    (b) $f(x, y) = x^2 - y^2$ for $k = 0$ and $k = 2$

    (c) $f(x_1, x_2, x_3, x_4) = x_1^2 + x_2^2 + x_3^2 + x_4^2$ for $k = 9$

Answers

1. Draw a vertical line (perpendicular to the z axis) and see if it intercepts the graph more than once. A given point in the domain (x, y) cannot have more than one associated z values

2. (a) $f : \mathbb{R}^3 \mapsto \mathbb{R}^5$

    (b) 8

    (c) 3

    (d) 5;3

3. (a) $F(x) = \langle x, x^3 + 5 \rangle$; $F(x, y) = y - x^3 - 5$

    (b) $F(x, y) = \langle x, y, x^2 + y^2 \rangle$; $F(x, y, z) = z - x^2 - y^2$

4. (a) Lines with slope 2 and y-intercept k

    (b) For $k = 0$, the lines $y = \pm x$; for $k = 2$, a hyperbola

(c) The collection of points in four dimensional space which are a distance of 3 from the origin. This can also be called a 3-sphere (a regular sphere is a 2-sphere because it is a two dimensional surface) or a hypersphere

2.3 Limits and Continuity

In this section we will apply the concept of a limit to multivariable functions. As in single variable calculus, limits are a tool we use to look at the behavior of a function around a certain input value.

For a single variable real valued function $f(x)$, the limit as $x \to a$ is the number that $f(x)$ approaches when x is really close to, but not directly at, a. The domain for such a function is one dimensional, so thinking of the domain as a number line, x can approach a from the right or left, corresponding to right and left hand limits. No matter which direction x approaches a from, the limit must be the same if it exists.

Limits for multivariable real valued functions are not that big of a stretch from this. Let's take a two variable function $f(x, y)$. The limit as $(x, y) \to (a_1, a_2)$ is the number that $f(x, y)$ approaches when (x, y) approaches the point (a_1, a_2) in the two dimensional domain of the function. Notice that in a two dimensional domain, there are a lot more ways by which (x, y) can approach (a_1, a_2) than just right or left. If the limit exists, it must have the same value for ALL of these possible paths.

Definition of a Limit

Throughout the rest of the section, we will represent points as vectors to tidy up the notation. For example, instead of talking about the limit as $(x, y) \to (a_1, a_2)$, we will say $\mathbf{x} \to \mathbf{a}$, where these vectors are understood to stand for the points in the first expression.

Suppose $f(\mathbf{x})$ is a real valued function. The expression

$$\lim_{\mathbf{x} \to \mathbf{a}} f(\mathbf{x}) = L$$

means that we can make $f(\mathbf{x})$ as close to L as we want by making $\mathbf{x}$ close enough to $\mathbf{a}$.

As a concrete example, let's take the single variable function $f(x) = x^2$. You know that $\lim_{x \to 2} f(x) = 4$. According to this definition, this means that we can make $f(x)$ as close to 4 as we want by making x close enough to 2. If I wanted to ensure that $f(x)$ is within a distance .5 from the limit 4, I could accomplish that by only looking at x values within a range of .1 around 2. For any x in the interval $1.9 < x < 2.1$, the corresponding output will be within the interval $3.5 < f(x) < 4.5$.

In a multivariable function, this "distance" between $f(\mathbf{x})$ and L and between $\mathbf{x}$ and $\mathbf{a}$ becomes the magnitude of a vector, and the "interval" around $\mathbf{a}$ becomes a disc or ball for a two or three dimensional domain, respectively. We can now tackle the formal definition of a limit, also referred to as the ***epsilon delta definition of a limit***.

Given a real valued function $f(\mathbf{x})$, we say that the limit of $f(\mathbf{x})$ as $\mathbf{x}$ approaches $\mathbf{a}$ equals L if:

- For any positive number ε there exists a corresponding positive number δ

- IF $0 < |\mathbf{x} - \mathbf{a}| < \delta$, THEN $|f(\mathbf{x}) - L| < \varepsilon$

Here ε just stands for how close we want $f(\mathbf{x})$ to be to L. For any such distance that we choose, we can find a number δ such that if we only look at points in the domain $\mathbf{x}$ within a distance δ from $\mathbf{a}$ (but not at $\mathbf{a}$ itself), the outputs of the function at these points will always be within the desired distance ε from the limit L.

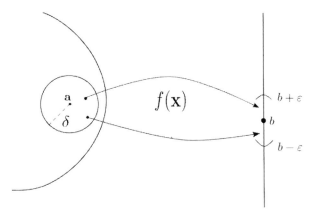

Figure 2.3: An illustration of a limit for a two variable real valued function. The function assigns any points within a distance δ from $\mathbf{a}$ to an output within a distance ε from b, the limit.

Let's look at an example with a two variable function $f(x, y) = x^2 + y^2$. You might guess intuitively that

$$\lim_{(x,y)\to(0,0)} x^2 + y^2 = 0$$

Can we back up this guess using the definition of a limit? First, suppose we choose $\varepsilon = 4$, so we want to find a δ such that if we only look at points within that distance from $(0,0)$, then $f(x, y)$ will always be within 4 of 0. Since $f(x, y)$ is always positive, that means we want $f(x, y) < 4$.

For this value of ε, the choice $\delta = 2$ fulfills the requirement because if we choose points in the domain within a distance of 2 from the origin,

$$0 < |\mathbf{x} - \mathbf{a}| < 2$$

$$0 < \sqrt{(x - 0)^2 + (y - 0)^2} < 2$$

$$0 < x^2 + y^2 < 4$$

$$f(x,y) < 4$$

In general, we could prove that for any ε the choice $\delta = \sqrt{\varepsilon}$ would work, but we will not cover limit proofs in this book.

Limits Along Different Paths

Often times it is easier to prove that a limit does not exist than to prove that it does exist, because all we have to do is show that the limit as $\mathbf{x}$ approaches $\mathbf{a}$ is not the same for ALL possible paths. This is analogous to showing that the left and right hand limits are different for a single variable function. As a first example, let's look at the behavior as $(x, y) \to (0, 0)$ of the function

$$f(x, y) = \frac{xy}{x^2 + y^2}$$

The domain of this function is the entire xy plane except the origin. Obviously, we cannot just plug in $(0, 0)$ to evaluate the limit in this case, so we will have to look at certain paths. By a path I mean a particular curve in the domain along which we will allow $\mathbf{x}$ to approach $\mathbf{0}$. To start with, we will look at the function's behavior as $\mathbf{x}$ approaches the origin along the x axis. Along the x axis, $y = 0$, so the function becomes

$$f(x, y) = \frac{0}{x^2} = 0$$

This expression is no longer undefined at $(0, 0)$, so we see that $\lim\limits_{\mathbf{x} \to \mathbf{0}} f(x, y) = 0$ when we approach along the x axis. If we approach the origin along the y axis, where $x = 0$, the function becomes 0 again so the limit is again 0. So far we've checked two paths and the limit has been 0 both times, but we cannot conclude that the limit is 0 yet. Let's approach the origin along the line $y = x$ in the domain. On this line, the function becomes

$$f(x, y) = \frac{x^2}{x^2 + x^2} = \frac{x^2}{2x^2} = \frac{1}{2}$$

So the limit along this curve is $\frac{1}{2}$, which is different than the limit along the x and y axes. Because we found a path along which the limit is different than the limit along another path, the limit of the function as $\mathbf{x} \to \mathbf{0}$ cannot exist.

To save some time, we might have let $\mathbf{x}$ approach the origin along a line with slope m, $y = mx$ which would give us

$$f(x, y) = \frac{mx^2}{x^2 + m^2 x^2} = \frac{m}{1 + m^2}$$

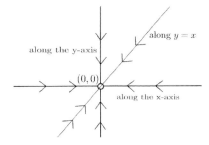

Figure 2.4: Letting $\mathbf{x} \to \mathbf{a}$ along three different paths

So we see that the limit is actually different for lines of different slopes. Even if we found that the limit was the same for all lines through the origin, we could then check paths such as $y = x^2$, $y = x^3$ and so on.

As you can tell from this example, proving that a limit *doesn't* exist is relatively simple: find a path by which $\mathbf{x} \to \mathbf{a}$ where the limit is different than on another path. However, keep in mind that doing this process will NOT prove that a limit *does* exist. You would need to check every possible path, which would be impossible.

Continuity

Fortunately, the limits of many functions can be evaluated by direct substitution, that is, just plugging in $\mathbf{a}$ to the function. To describe when this is true, we need the concept of **continuity**. Just as in the single variable case, a function is said to be continuous at a point in the domain $\mathbf{a}$ if the limit there and the actual value of the function both exist and are equal.

$$\lim_{\mathbf{x} \to \mathbf{a}} f(\mathbf{x}) = f(\mathbf{a})$$

A function is said to be a continuous function if it is continuous at every point *in its domain*. To illustrate this point, consider the function $f(x) = \frac{2x}{x-2}$, which has a vertical asymptote at $x = 2$. We consider this to be a continuous function; the discontinuity at $x = 2$ is irrelevant because it is not in the domain of the function.

Many types of functions are always continuous, such as polynomials, trig functions, logs, and exponential functions. In addition, if we have two continuous functions f and g, then the following combinations are also continuous:

- The sum $f + g$

- The product fg

- A constant multiple cf

- The quotient f/g (if $g = 0$ at some point, then that point is not in the domain of f/g)

- The composition $f \circ g$

The definition of continuity implies that if the function $f(\mathbf{x})$ is continuous at $\mathbf{a}$, then we can evaluate the limit by direct substitution.

Example. Find the limit

$$\lim_{(x,y,z)\to(1,2,3)} \frac{x^2 + zy^3}{xy + e^{2z}}$$

The top is a polynomial, which has to be continuous. The bottom is the sum of a polynomial and an exponential function, which are both continuous. The bottom is also nonzero at this point, so we can compute the limit by direct substitution.

$$\lim_{(x,y,z)\to(1,2,3)} \frac{x^2 + zy^3}{xy + e^{2z}} = \frac{1 + (3)(2)^3}{(1)(2) + e^{2(3)}} = \frac{25}{2 + e^6}$$

Example. Find the limit

$$\lim_{(x,y)\to(1,2)} \sin(\ln(x^2 + y^3))$$

This is the composition of three functions: a trig function, log, and polynomial. All of these are continuous at the points in question (we have to be careful that the log will not get a negative or zero input), so we can compute by direct substitution.

$$\lim_{(x,y)\to(1,2)} \sin(\ln(x^2 + y^3)) = \sin(\ln(9))$$

Vector Valued Functions

Lastly, we need to discuss limits for vector functions. Everything we said about limits for real valued functions is still true; the only difference is that the limit of a vector valued function will be a vector instead of a number.

When computing the limit of a vector function, we can instead compute the limits of each coordinate function individually. We say that limits can be taken *coordinatewise*.

Example. Find the limit

$$\lim_{t\to 2}\langle 2t, t^3 - 1, e^{4t}\rangle$$

Each of the coordinate functions is continuous at $t = 2$, so we can compute coordinatewise by direct substitution.

$$\lim_{t\to 2}\langle 2t, t^3 - 1, e^{4t}\rangle = \langle \lim_{t\to 2} 2t, \lim_{t\to 2} t^3 - 1, \lim_{t\to 2} e^{4t}\rangle$$

$$= \langle 4, 7, e^8\rangle$$

Exercises

1. Prove that the limit does not exist

$$\lim_{\mathbf{x}\to\mathbf{0}} \frac{x^2 y^2}{x^4 + 3y^4}$$

2. Prove that the limit does not exist

$$\lim_{\mathbf{x}\to\mathbf{0}} \frac{x^3 y}{x^7 + y^2}$$

3. Make sure that you can use direct substitution, then evaluate the limit

   (a)

   $$\lim_{(x,y)\to(1,2)} \frac{x^3 y^2 + \sin(xy)}{y^5 + e^{2x}}$$

   (b)

   $$\lim_{t\to 3} \langle 3t^2, \ln(\frac{t}{3}) \rangle$$

4. Sometimes we can use a substitution to convert a multivariable limit to a single variable limit. Evaluate the following limits by converting to polar coordinates, using the fact that $(x, y) \to (0, 0)$ is the same condition as $r \to 0$ (this *only* works for limits at $(0, 0)$ because otherwise it would involve θ)

   (a)

   $$\lim_{\mathbf{x}\to\mathbf{0}} \frac{x^3 y^2}{x^2 + y^2}$$

   (b)

   $$\lim_{\mathbf{x}\to\mathbf{0}} \frac{1 - \cos(x^2 + y^2)}{x^2 + y^2}$$

Answers

1. An example path that is different would be $y = x$

2. An example path that is different would be $y = x^3$

3. (a) $\frac{4 + \sin(2)}{32 + e^2}$

   (b) $\langle 27, 0 \rangle$

4. (a) $\lim_{r\to 0} \frac{r^5 \cos^3\theta \sin^2\theta}{r^2} = \lim_{r\to 0} r^3 \cos^3\theta \sin^2\theta = 0$

   (b) $\lim_{r\to 0} \frac{1 - \cos r^2}{r^2}$. Use L'Hopital's Rule to get $\lim_{r\to 0} \frac{2r \sin r^2}{2r} = \lim_{r\to 0} \sin r^2 = 0$

2.4 Linear Transformations

In this section we will study a special type of function called a ***linear transformation***. The big idea is that linear functions (such as lines and planes) are easier to work with than nonlinear ones and have some nice properties, which we will see shortly. They will also pop up when we look at differentials of multivariable functions.

A linear transformation is any function which has the following two properties:

$$L(c\mathbf{x}) = cL(\mathbf{x})$$

$$L(\mathbf{x} + \mathbf{y}) = L(\mathbf{x}) + L(\mathbf{y})$$

Here c is a constant while $\mathbf{x}$ and $\mathbf{y}$ are input vectors of the function. So basically, a linear transformation is a function where you can move a constant out and split up a sum.

Example. The function $L : \mathbb{R}^2 \mapsto \mathbb{R}$ given by $L(\mathbf{x}) = x_1 + x_2$ is a linear transformation. To verify that any function is a linear transformation, we simply check that it satisfies the two properties.

$$L(c\mathbf{x}) = cx_1 + cx_2 = c(x_1 + x_2) = cL(\mathbf{x})$$

$$L(\mathbf{x} + \mathbf{y}) = (x_1 + y_1) + (x_2 + y_2) = (x_1 + x_2) + (y_1 + y_2) = L(\mathbf{x}) + L(\mathbf{y})$$

Example. From single variable calculus you know that the differential of a function $f(x)$ at a point a is $df_a = f'(a)(x - a)$. If we let h stand for $(x - a)$, then we can write $df_a(h) = f'(a)h$. This is a linear transformation $df_a : \mathbb{R} \mapsto \mathbb{R}$.

$$df_a(ch) = f'(a)ch = c(f'(a)h) = cdf_a(h)$$

$$df_a(h_1 + h_2) = f'(a)(h_1 + h_2) = f'(a)h_1 + f'(a)h_2 = df_a(h_1) + df_a(h_2)$$

Theorem. For any linear transformation, $L(\mathbf{0}) = \mathbf{0}$.

Proof. To prove this, we just have to choose $c = 0$ in the first property of linear transformations.

$$L(0\mathbf{x}) = 0L(\mathbf{x})$$

$$L(\mathbf{0}) = \mathbf{0}$$

□

One of the most important properties of linear transformations is that every linear transformation has an associated matrix, and the function can then be written as the product of this matrix and the input vector. If we call this matrix A, then $L(\mathbf{x}) = A\mathbf{x}$, where $\mathbf{x}$ is the input vector represented as a column matrix. To see what this means, consider the following example:

$$L(\mathbf{x}) = \begin{bmatrix} 3x_1 + x_2 \\ 2x_1 - x_2 \\ 5x_1 \end{bmatrix}$$

You can verify that this is a linear transformation by checking the two properties as in the first two examples. This is a transformation $L : \mathbb{R}^2 \mapsto \mathbb{R}^3$, so it will have a corresponding 3×2 matrix.

$$A = \begin{bmatrix} 3 & 1 \\ 2 & -1 \\ 5 & 0 \end{bmatrix}$$

You can check that this is true by multiplying the matrices A and $\mathbf{x}$ to get the original function.

$$A\mathbf{x} = \begin{bmatrix} 3 & 1 \\ 2 & -1 \\ 5 & 0 \end{bmatrix} \begin{bmatrix} x_1 \\ x_2 \end{bmatrix} = \begin{bmatrix} 3x_1 + x_2 \\ 2x_1 - x_2 \\ 5x_1 \end{bmatrix}$$

Notice that A is 3×2 and $\mathbf{x}$ is 2×1 so the multiplication is possible and yields a 3×1 matrix, which is the dimension of the linear transformation's output vector.

It turns out that we also have a convenient way of finding the matrix for any linear transformation. This method utilizes the fact that any vector can be written as a combination of the standard basis vectors. In our current example, the domain is two dimensional, so any input vector $\mathbf{x} = \langle x_1, x_2 \rangle$ can be written as $x_1\mathbf{e}_1 + x_2\mathbf{e}_2$. By the properties of linear transformations,

$$L(\mathbf{x}) = L(x_1\mathbf{e}_1 + x_2\mathbf{e}_2) = x_1 L(\mathbf{e}_1) + x_2 L(\mathbf{e}_2)$$

The values of $L(\mathbf{e}_2)$ and $L(\mathbf{e}_2)$ can be easily computed by just plugging them into the function.

$$L(\mathbf{e}_1) = \begin{bmatrix} 3 \\ 2 \\ 5 \end{bmatrix} \quad \text{and} \quad L(\mathbf{e}_2) = \begin{bmatrix} 1 \\ -1 \\ 0 \end{bmatrix}$$

So the function can be written like

$$L(\mathbf{x}) = x_1 L(\mathbf{e}_1) + x_2 L(\mathbf{e}_2)$$

$$= x_1 \begin{bmatrix} 3 \\ 2 \\ 5 \end{bmatrix} + x_2 \begin{bmatrix} 1 \\ -1 \\ 0 \end{bmatrix}$$

$$= \begin{bmatrix} 3 & 1 \\ 2 & -1 \\ 5 & 0 \end{bmatrix} \begin{bmatrix} x_1 \\ x_2 \end{bmatrix}$$

We see that plugging the standard basis vectors into the linear transform give us the columns of its corresponding matrix! The function evaluated at the first standard basis vector $\mathbf{e}_1$ gave us the first column of the matrix A, and the second standard basis vector gave us the second column. These ideas are true for any linear transformation.

Theorem. For any linear transformation $L : \mathbb{R}^n \mapsto \mathbb{R}^m$, there is a corresponding $m \times n$ matrix A such that

$$L(\mathbf{x}) = A\mathbf{x}$$

and the ith column of A can be obtained by plugging the ith standard basis vector into the function.

Example. For a linear transform $L : \mathbb{R}^3 \mapsto \mathbb{R}$, which is a real valued function, the corresponding matrix will be of dimension 1×3, a row vector.

$$A = \begin{bmatrix} a_1 & a_2 & a_3 \end{bmatrix}$$

Therefore,

$$L(\mathbf{x}) = A\mathbf{x} = \begin{bmatrix} a_1 & a_2 & a_3 \end{bmatrix} \begin{bmatrix} x_1 \\ x_2 \\ x_3 \end{bmatrix} = a_1 x_1 + a_2 x_2 + a_3 x_3$$

Notice that this can also be represented by the dot product, if $\mathbf{A}$ is thought of as a vector

$$L(\mathbf{x}) = \mathbf{A} \bullet \mathbf{x}$$

To end the chapter, we look at what happens when we have a composition of linear transforms. Let $L_1(\mathbf{x})$ be the linear transform we were working with before, whose matrix A_1 was

$$A_1 = \begin{bmatrix} 3 & 1 \\ 2 & -1 \\ 5 & 0 \end{bmatrix}$$

Now introduce a new linear transform $L : \mathbb{R}^3 \mapsto \mathbb{R}$ given by $L(\mathbf{x}) = x_1 + x_2 + x_3$. The matrix of this linear transform is

$$A_2 = \begin{bmatrix} 1 & 1 & 1 \end{bmatrix}$$

What does the composition $L_2 \circ L_1$ look like? This will be a transform $\mathbb{R}^2 \mapsto \mathbb{R}$ and therefore should have a 1×2 matrix associated with it.

$$L_2(L_1(\mathbf{x})) = (3x_1 + x_2) + (2x_1 - x_2) + (5x_1) = 10x_1$$

So it looks like the matrix of the composition transform is

$$A = \begin{bmatrix} 10 & 0 \end{bmatrix}$$

Now notice that this new matrix is the product of the two matrices A_2 and A_1 of the original linear transforms involved.

$$A_2 A_1 = \begin{bmatrix} 1 & 1 & 1 \end{bmatrix} \begin{bmatrix} 3 & 1 \\ 2 & -1 \\ 5 & 0 \end{bmatrix} = \begin{bmatrix} 10 & 0 \end{bmatrix}$$

In general, if we have a composition of linear transforms, to get the matrix of the new transform we just multiply the matrices of the original transforms. We can see this intuitively like so

$$L_1(\mathbf{x}) = A_1 \mathbf{x}$$
$$L_2(\mathbf{x}) = A_2 \mathbf{x}$$
$$L_2 \circ L_1(\mathbf{x}) = A_2(A_1 \mathbf{x}) = (A_2 A_1)\mathbf{x}$$

Exercises

1. Show that the line $f(x) = mx + b$ is linear only if it passes through the origin, i.e. $b = 0$

2. Show that if the single condition

$$L(a\mathbf{x} + b\mathbf{y}) = aL(\mathbf{x}) + bL(\mathbf{y})$$

is true for a function, then it implies that the two properties of linear transforms are true. (a and b are just constants)

3. Show that the following function is a linear transform and find its matrix

$$L(\mathbf{x}) = \begin{bmatrix} 2x_1 \\ x_1 + 3x_2 \\ 4x_1 + x_2 \end{bmatrix}$$

4. Find the matrix of the linear transform by plugging in the standard basis vectors

$$L(\mathbf{x}) = \begin{bmatrix} 3x_1 + x_2 \\ x_1 - 3x_2 \\ 5x_2 \end{bmatrix}$$

5. Let $R_\theta(\mathbf{x})$ be a linear transform $R_\theta : \mathbb{R}^2 \mapsto \mathbb{R}^2$ which takes a point (x_1, x_2) and rotates it counterclockwise through the angle θ.

   (a) Find the matrix of this transform (Hint: draw the standard basis vectors and use trigonometry to find their new positions after a rotation)

   (b) Performing one rotation through 2θ is the same thing as performing two rotations through θ. Since doing two separate rotations is like taking a composition, this is the same thing as saying

   $$R_{2\theta}(\mathbf{x}) = R_\theta(\mathbf{x}) \circ R_\theta(\mathbf{x})$$

   Use this to prove the double sine and cosine identities

   (c) Performing one rotation through an angle $\theta + \alpha$ is equal to performing two separate rotations through α and then θ. Use a similar method as in part (b) to prove the sine and cosine addition identities

Answers

1. If $b \neq 0$, it cannot be linear because

$$f(cx) = m(cx) + b = cmx + b \neq cf(x)$$

If it does pass through the origin, then it is linear because

$$f(cx) = m(cx) = m(cx) = cf(x)$$

$$f(x + y) = m(x + y) = mx + my = f(x) + f(y)$$

2. For the first property, choose either $a = 0$ or $b = 0$; for the second property, choose $a = b = 1$

3. Using the single condition from exercise 2,

$$L(a\mathbf{x} + b\mathbf{y}) = \begin{bmatrix} 2(ax_1 + by_1) \\ (ax_1 + by_1) + 3(ax_2 + by_2) \\ 4(ax_1 + by_1) + (ax_2 + by_2) \end{bmatrix}$$

$$= \begin{bmatrix} 2ax_1 \\ ax_1 + 3ax_2 \\ 4ax_1 + ax_2 \end{bmatrix} + \begin{bmatrix} 2by_1 \\ by_1 + 3by_2 \\ 4by_1 + by_2 \end{bmatrix}$$

$$= aL(\mathbf{x}) + bL(\mathbf{y})$$

Its matrix is

$$A = \begin{bmatrix} 2 & 0 \\ 1 & 3 \\ 4 & 1 \end{bmatrix}$$

4.

$$L(\mathbf{e}_1) = \begin{bmatrix} 3 \\ 1 \\ 0 \end{bmatrix} \qquad L(\mathbf{e}_2) = \begin{bmatrix} 1 \\ -3 \\ 5 \end{bmatrix}$$

$$A = \begin{bmatrix} 3 & 1 \\ 1 & -3 \\ 0 & 5 \end{bmatrix}$$

5. (a)

$$A = \begin{bmatrix} \cos\theta & -\sin\theta \\ \sin\theta & \cos\theta \end{bmatrix}$$

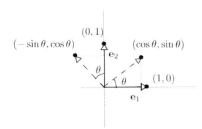

(b) Look at the matrices of both sides of the equation

$$\begin{bmatrix} \cos 2\theta & -\sin 2\theta \\ \sin 2\theta & \cos 2\theta \end{bmatrix} = \begin{bmatrix} \cos\theta & -\sin\theta \\ \sin\theta & \cos\theta \end{bmatrix}\begin{bmatrix} \cos\theta & -\sin\theta \\ \sin\theta & \cos\theta \end{bmatrix}$$

$$\begin{bmatrix} \cos 2\theta & -\sin 2\theta \\ \sin 2\theta & \cos 2\theta \end{bmatrix} = \begin{bmatrix} \cos^2\theta - \sin^2\theta & -2\sin\theta\cos\theta \\ 2\sin\theta\cos\theta & \cos^2\theta - \sin^2\theta \end{bmatrix}$$

(c) We have

$$R_{\theta+\alpha}(\mathbf{x}) = R_\theta(\mathbf{x}) \circ R_\alpha(\mathbf{x})$$

Looking at the matrices on both sides,

$$\begin{bmatrix} \cos(\theta+\alpha) & -\sin(\theta+\alpha) \\ \sin(\theta+\alpha) & \cos(\theta+\alpha) \end{bmatrix} = \begin{bmatrix} \cos\theta & -\sin\theta \\ \sin\theta & \cos\theta \end{bmatrix} \begin{bmatrix} \cos\alpha & -\sin\alpha \\ \sin\alpha & \cos\alpha \end{bmatrix}$$

$$\begin{bmatrix} \cos(\theta+\alpha) & -\sin(\theta+\alpha) \\ \sin(\theta+\alpha) & \cos(\theta+\alpha) \end{bmatrix} = \begin{bmatrix} \cos\theta\cos\alpha - \sin\theta\sin\alpha & -\sin\theta\cos\alpha - \cos\theta\sin\alpha \\ \sin\theta\cos\alpha + \cos\theta\sin\alpha & \cos\theta\cos\alpha - \sin\theta\sin\alpha \end{bmatrix}$$

Chapter 3

Differential Calculus

3.1   Parametric Curves

Before we look at derivatives of general multivariable functions, we will first see how the ideas of differential calculus apply to single variable vector valued functions. In general, the image of such a function will be a *curve* in two or three dimensional space, depending on the dimensions of the function's range.

In the last chapter, we already saw how the image of

$$\gamma(t) = \langle t, t^2 \rangle$$

was the parabola $y = x^2$ in two dimensional space. We say that the function $\gamma(t)$ is a *parametrization* of the parabola. In general, a function is said to be a parametrization of a curve C if the function's image matches the curve. Throughout this book, we will use the notation $\gamma(t)$ to represent parametric curves.

One way to think about why single variable vector functions generate curves is to remember that the dimensions of a function's domain usually determine the dimensions of the object it represent (not always, e.g. level sets). Therefore, single variable functions will generate one dimensional objects, which are curves. The graph of a two variable real valued function $f(x, y)$ is a two dimensional surface in three dimensional space. Another way to look at it is to imagine the t number line (the domain) being lifted up and thrown into two or three dimensional space in some configuration which resembles a curve.

When finding a parametrization of a curve, we must keep in mind that the *parameter* t has some sort of meaning; it stands for something. In the parabola above, t is really standing for the x coordinate of each point because the first coordinate function is just t.

As another example, consider the common parametrization of the unit circle

$$\gamma(t) = \langle \cos t, \sin t \rangle$$

Here t represents the angle θ from polar coordinates. In fact, to remind ourselves of this we could write the curve as

$$\gamma(\theta) = \langle \cos\theta, \sin\theta \rangle$$

There is no rule saying we have to use the letter t in our parametrizations.

A common three dimensional curve is the helix, which can be parametrized like so

$$\gamma(t) = \langle \cos t, \sin t, t \rangle$$

The x and y components go around in the unit circle while the z component steadily climbs higher, resulting in an upward spiral.

Keep in mind that the outputs of these functions are still *vectors*; they are vector functions, after all. However, we interpret these vectors as the position vector for a point on the curve.

Derivatives

We will look at parametric curves again in chapter 5; the main thing we need to cover here is the derivative of these vector functions. For example, if $\gamma(t)$ represented the position vector of some moving particle at a time t, then the derivative $\gamma'(t)$ would represent the particle's velocity vector at that time.

Since these are still single variable functions, the expression for the derivative looks exactly the same as the one you're used to. The derivative of a function $\gamma(t)$ at the point $t = a$ is

$$\gamma'(a) = \lim_{h \to 0} \frac{\gamma(a+h) - \gamma(a)}{h}$$

The only difference is that this is now a vector equation. If we name the coordinate functions of $\gamma(t)$ as $x(t)$, $y(t)$, and $z(t)$, so $\gamma(t) = \langle x(t), y(t), z(t) \rangle$, then the above expression is really saying

$$\gamma'(a) = \lim_{h \to 0} \langle \frac{x(a+h) - x(a)}{h}, \frac{y(a+h) - y(a)}{h}, \frac{z(a+h) - z(a)}{h} \rangle$$

But in chapter 2 we learned that the limits of vector valued functions can be computed coordinatewise, i.e. we can take the limit of each component separately. This suggests that derivatives can also be computed coordinatewise, so that

$$\gamma'(a) = \langle x'(a), y'(a), z'(a) \rangle$$

Example. Find the derivative of

$$\gamma(t) = \langle 3t^2, \sin t, e^{4t} \rangle$$

All we do is compute the derivatives of the coordinate functions individually.

$$\gamma'(t) = \langle 6t, \cos t, 4e^{4t} \rangle$$

For a single variable function, the derivative signified the slope of the tangent line at the point where the derivative was evaluated. Since the derivative of a vector function is a vector instead of a number, this relationship obviously cannot still be true. Can we assign some sort of meaning to the derivative of a vector function $\gamma'(a)$?

To answer this question, we need to interpret the definition of the derivative geometrically. $\gamma(a)$ is the position vector of the point on the curve corresponding to $t = a$. $\gamma(a + h)$ then represents a point a little further down the curve. The difference vector $\gamma(a + h) - \gamma(a)$ is therefore a vector which points from the point $\gamma(a)$ to the point down the curve $\gamma(a + h)$.

As $h \to 0$, the points come closer together on the curve, and we can see that the difference vector will point *tangent* to the curve at $\gamma(a)$. In other words, the derivative of a vector function at $t = a$ can be visualized as a tangent vector to the curve at the point $\gamma(a)$.

Exercises

1. Find the derivative $\gamma(t) = \langle e^{\sin(t^2)}, \ln(t^5 - 2t^2), t^t \rangle$

2. Prove that the product rule holds for the derivative of the dot product of two vector functions

$$\frac{d}{dt}(\gamma_1(t) \cdot \gamma_2(t)) = \gamma_1'(t) \cdot \gamma_2(t) + \gamma_1(t) \cdot \gamma_2'(t)$$

3. Use the product rule to prove that if a vector function $\gamma(t)$ has a *constant magnitude*, then it is always orthogonal to its derivative vector $\gamma'(t)$ (Hint: if the magnitude is constant, then $\frac{d}{dt}|\gamma(t)|^2 = 0$)

4. Let $\gamma(t)$ be the position vector of a particle moving with constant acceleration vector $\gamma''(t) = \mathbf{a}$. Find the position vector if the initial velocity and position are $\gamma'(0) = \mathbf{v}_0$ and $\gamma(0) = \mathbf{r}_0$.

   (a) Describe the particle's path if $\mathbf{a} = 0$

5. Consider a particle moving with position vector $\gamma(t) = \langle 2\cos t, 2\sin t, 3t \rangle$

   (a) Describe the path of the particle

   (b) Show that the particle has a constant speed (speed is the magnitude of velocity)

   (c) Show that the particle's derivative vector makes a constant nonzero angle with the z axis

   (d) Given $t_1 = 0$ and $t_2 = 2\pi$, find the vector $\gamma(t_2) - \gamma(t_1)$ and show that it is vertical

   (e) Conclude that the equation $\gamma(t_2) - \gamma(t_1) = (t_2 - t_1)\gamma'(c)$ cannot be true for any value of c between t_1 and t_2. This means that the mean value theorem does not hold for vector valued functions

Answers

1. $\gamma'(t) = \langle 2t\cos(t^2)e^{\sin(t^2)}, \frac{5t^4-4t}{t^5-2t^2}, t^t(\ln t + 1) \rangle$

2. Write out the dot product directly, then differentiate. Since the coordinate functions are all single variable functions we know the product rule works on them.

$$\frac{d}{dt}(\gamma_1(t) \cdot \gamma_2(t)) = \frac{d}{dt}(x_1(t)x_2(t) + y_1(t)y_2(t) + z_1(t)z_2(t))$$
$$= (x_1'(t)x_2(t) + x_1(t)x_2'(t)) + (y_1'(t)y_2(t) + y_1(t)y_2'(t))$$
$$+ (z_1'(t)z_2(t) + z_1(t)z_2'(t))$$
$$= (x_1'(t)x_2(t) + y_1'(t)y_2(t) + z_1'(t)z_2(t)) + (x_1(t)x_2'(t) + y_1(t)y_2'(t) + z_1(t)z_2'(t))$$
$$= \gamma_1'(t) \cdot \gamma_2(t) + \gamma_1(t) \cdot \gamma_2'(t)$$

3.

$$\frac{d}{dt}|\gamma(t)|^2 = \frac{d}{dt}(\gamma(t) \cdot \gamma(t))$$
$$= \gamma'(t) \cdot \gamma(t) + \gamma(t) \cdot \gamma'(t)$$
$$= 2(\gamma'(t) \cdot \gamma(t)) = 0$$

4. $\gamma(t) = \frac{1}{2}t^2\mathbf{a} + t\mathbf{v}_0 + \mathbf{r}_0$

(a) a straight line through $\mathbf{r}_0$ with velocity vector $\mathbf{v}_0$

5. (a) a helix of radius 2

   (b) $|\gamma'(t)| = \sqrt{13}$

   (c) $\theta = \arccos \frac{3}{\sqrt{13}}$

   (d) $\gamma(t_2) - \gamma(t_1) = \langle 0, 0, 6\pi \rangle$

   (e) $\gamma(t_2) - \gamma(t_1)$ is a vertical vector, but in part (c) we found that the derivative vector is never parallel to the z axis, so therefore these two vectors cannot be scalar multiples of each other, and $(t_2 - t_1)$ is just a scalar

3.2 Directional Derivatives

We now want to extend the ideas of differential calculus to more general multivariable functions. The most important tool from differential calculus is of course the derivative, which tells us how much the output of a function is changing in response to changes in the input. In a plain old single variable function $y = f(x)$, or in the single variable vector functions we looked at last section, the domain is one dimensional. So when we talk about changes in the input, the only things possible are a movement right or left along the domain, which is just a number line. The derivative $f'(a)$ tells us: if we start at a and move 1 to the right in the domain, what is the corresponding (approximate) change in the output $f(x)$?

For a multivariable function, there are many more ways to move about in the domain. If we had a two variable real valued function $f : \mathbb{R}^2 \mapsto \mathbb{R}$, we could still ask: what change does a movement right in the domain cause in the output? For example, we could ask how the function changes when we move along the vector $\langle 5, 0 \rangle$ in the domain. But now that we have a two dimensional domain, we can move in so many more directions. We could ask: what is the change in the function when we start at a point $\mathbf{a}$ and move along the vector $\langle 3, 4 \rangle$? How about $\langle -4, 7 \rangle$? To answer these questions, we need a new type of derivative: the ***directional derivative***.

When we take the directional derivative of a function, we must have a certain direction in mind. The vector $\mathbf{v}$ will be this direction that we are moving in the domain. We also must specify a point in the domain $\mathbf{a}$ where we start. So the directional derivative answers the question: if we start at $\mathbf{a}$ in the domain and move along the line given by direction $\mathbf{v}$, how is the output of the function changing?

Our path in the domain is represented by the equation of the line $x(t) = \mathbf{a} + t\mathbf{v}$. Since we want to know the behavior of the function along this line in the domain, we plug this line into the function.

$$\gamma(t) = f(\mathbf{a} + t\mathbf{v})$$

This new function $\gamma(t)$ represents whatever the output of the original function f is doing as we move along this particular line in the domain. Geometrically, the image of $\gamma(t)$ is the curve on the image of $f(\mathbf{x})$ which contains points corresponding to inputs along the line in the domain. For example, if the image of $f(\mathbf{x})$ is some surface in three dimensional space, then $\gamma(t)$ is a single curve which lies in the surface.

Next we take the derivative of this expression with respect to t at $t = 0$, $\gamma'(0)$

$$\gamma'(0) = \lim_{h \to 0} \frac{\gamma(h) - \gamma(0)}{h} = \lim_{h \to 0} \frac{f(\mathbf{a} + h\mathbf{v}) - f(\mathbf{a})}{h}$$

We call this the directional derivative of f with respect to $\mathbf{v}$ at $\mathbf{a}$, denoted by $D_{\mathbf{v}}f(\mathbf{a})$. This operation works for both real valued and vector valued functions. For a real valued function, the directional derivative will be a scalar, and for a vector valued function, the directional derivative will be a vector whose components are the directional derivatives of the coordinate functions.

Example. The temperature at a point (x, y) on a piece of paper is given by the real valued function $f(x, y) = x^2 + y^2$. If we walk on this piece of paper starting at the point $(1, 2)$ and moving in the direction $\langle 3, 5 \rangle$, what is the rate of change of temperature we experience as we pass through $(1, 2)$?

The quantity we are after is the directional derivative with respect to $\mathbf{v} = \langle 3, 5 \rangle$ at $\mathbf{a} = \langle 1, 2 \rangle$. Our movement in the domain causes a change in the output of the function, which in this case is a change in temperature. We are moving in the domain along the line

$$x(t) = \langle 1, 2 \rangle + t \langle 3, 5 \rangle = \langle 1 + 3t, 2 + 5t \rangle$$

The output of the function at points along this line is given by

$$\gamma(t) = f(\mathbf{a} + t\mathbf{v}) = f(1 + 3t, 2 + 5t) = (1 + 3t)^2 + (2 + 5t)^2$$

To get the directional derivative, differentiate this with respect to t and set $t = 0$

$$D_{\mathbf{v}}f(\mathbf{a}) = \gamma'(0) = 6(1 + 3t) + 10(2 + 5t) = 6 + 20 = 26$$

So if we start at $(1, 2)$ on the paper and move in the direction of the vector $\langle 3, 5 \rangle$, i.e. if this vector starts at $\mathbf{a}$ we walk to the tip of the vector, the temperature will change by approximately 26.

Unit Directional Derivatives

Some people like to say that the operation we just saw is only a directional derivative if $\mathbf{v}$ is a unit vector. In other words, you can only take directional derivatives with respect to a unit vector. In this book, we let $\mathbf{v}$ be whatever we want, but it is worth looking at directional derivatives with unit vectors. Of course, the process for computing the directional derivative does not suddenly change if $\mathbf{v}$ is a unit vector.

Consider the directional derivative with respect to a random not unit vector $D_{\mathbf{v}}f(\mathbf{a})$. Now suppose that $\mathbf{u}$ is a unit vector in the same direction as $\mathbf{v}$, so that $\mathbf{v} = |\mathbf{v}|\,\mathbf{u}$. Are the directional derivatives with respect to $\mathbf{v}$ and with respect to $\mathbf{u}$ related? If so, how? The answer will follow immediately from the following fact.

Theorem. Consider the directional derivative of f with respect to $\mathbf{v}$ at $\mathbf{a}$ $D_{\mathbf{v}}f(\mathbf{a})$. If we multiply the vector $\mathbf{v}$ by a constant k, it is the same thing as multiplying the value of the directional derivative by k.

$$D_{k\mathbf{v}}f(\mathbf{a}) = kD_{\mathbf{v}}f(\mathbf{a})$$

Proof. We prove this by computing $D_{k\mathbf{v}}f(\mathbf{a})$ directly from the definition of a directional derivative. We move along the line $x(t) = \mathbf{a} + t(k\mathbf{v})$ in the domain, which translates under the function to the curve

$$\gamma(t) = f(\mathbf{a} + t(k\mathbf{v}))$$

The directional derivative is the derivative of this curve at $t = 0$

$$\gamma'(0) = \lim_{h \to 0} \frac{\gamma(h) - \gamma(0)}{h} = \lim_{h \to 0} \frac{f(\mathbf{a} + kh\mathbf{v}) - f(\mathbf{a})}{h}$$

Now we will introduce a new constant $c = kh$. Notice that if $h \to 0$, then $c \to 0$. Plugging this in, the expression becomes

$$\lim_{c \to 0} \frac{f(\mathbf{a} + c\mathbf{v}) - f(\mathbf{a})}{c/k} = k \lim_{c \to 0} \frac{f(\mathbf{a} + c\mathbf{v}) - f(\mathbf{a})}{c} = kD_{\mathbf{v}}f(\mathbf{a})$$

In the last step, we recognized the limit as the definition of the directional derivative with respect to $\mathbf{v}$. It just has a c instead of an h like we had earlier. $\square$

So we have our answer to the relationship between the directional derivatives with respect to $\mathbf{v}$ and a unit vector in the same direction $\mathbf{u}$. If $\mathbf{v} = |\mathbf{v}|\,\mathbf{u}$, the theorem says that

$$D_{\mathbf{v}}f(\mathbf{a}) = |\mathbf{v}|\,D_{\mathbf{u}}f(\mathbf{a})$$

The unit directional derivative has a special geometrical meaning if we consider the graph of a real valued function $f(x, y)$. This graph will be a surface in three dimensional space. Imagine you are standing on the point $f(\mathbf{a})$ and start walking in the direction of $\mathbf{u}$. You are walking along a curve, and this curve has a tangent line at $f(\mathbf{a})$ in the direction of $\mathbf{u}$. The unit directional derivative will be the slope of this tangent line. This is like how the single variable derivative $f'(a)$ is the slope of the tangent line to the graph of f at $f(a)$. The slope is the change in height when you move a distance of 1 in the direction of the line. Since a unit vector has a length of 1, it makes sense that the unit directional derivative can represent slope.

Example. Consider the graph of $f(x, y) = x^2 - y^2$. If you start at the point on the graph corresponding to the input $(1, 1)$ and move in the direction of $\langle 2, 3 \rangle$, what is the initial slope of your path?

Because we want slope of the tangent line to the graph, we need to find the directional derivative with respect to a unit vector which points in the same direction as $\mathbf{v} = \langle 2, 3 \rangle$. We could either convert $\mathbf{v}$ to a unit vector and compute the directional derivative, or we could just compute it with respect to $\mathbf{v}$, then divide by $|\mathbf{v}|$ because of the relationship we found earlier. The line in the domain we are moving along is

$$x(t) = \langle 1, 1 \rangle + t\langle 2, 3 \rangle = \langle 1 + 2t, 1 + 3t \rangle$$

The image under the function of this line is

$$f(1 + 2t, 1 + 3t) = (1 + 2t)^2 - (1 + 3t)^2$$

Now take the derivative with respect to t at $t = 0$ to get the directional derivative

$$D_\mathbf{v} f(\mathbf{a}) = 4(1 + 2t) - 6(1 + 3t) = 4 - 6 = -2$$

To get the unit directional derivative, we divide by $|\mathbf{v}| = \sqrt{4 + 9} = \sqrt{13}$

$$D_\mathbf{u} f(\mathbf{a}) = \frac{-2}{\sqrt{13}}$$

Partial Derivatives

A very widely used special case of directional derivatives is **partial derivatives**. The partial derivatives of a function are just the directional derivatives with respect to the standard basis vectors.

The directional derivative at $\mathbf{a}$ with respect to $\mathbf{e}_1$ is written as $D_1 f(\mathbf{a})$ and represents the change in the function when we move a distance of 1 in the x direction. The directional derivative with respect to $\mathbf{e}_2$ is written as $D_2 f(\mathbf{a})$, and so on.

Partial derivatives also have their own special notation similar to the Leibniz notation for single variable derivatives. $D_1 f(\mathbf{x})$ is referred to as the partial derivative with respect to x, $D_2 f(\mathbf{x})$ the partial with respect to y, etc.

$$D_{\mathbf{e_1}} f(\mathbf{x}) = D_1 f(\mathbf{x}) = \frac{\partial f}{\partial x}$$

$$D_{\mathbf{e_2}} f(\mathbf{x}) = D_2 f(\mathbf{x}) = \frac{\partial f}{\partial y}$$

$$D_{\mathbf{e_3}} f(\mathbf{x}) = D_3 f(\mathbf{x}) = \frac{\partial f}{\partial z}$$

Computing Partial Derivatives

The nice thing about partial derivatives is that they are very easy to compute. We show this by first using the definition of the directional derivative to compute a partial. Consider the direction derivative of $f(x, y)$ with respect to the first standard basis vector $\mathbf{v} = \langle 1, 0 \rangle$ at some point (a_1, a_2). The line through the domain we move along is

$$x(t) = \langle a_1 + t, a_2 \rangle$$

Then we have

$$\gamma(t) = f(a_1 + t, a_2)$$

$$\gamma'(0) = \lim_{h \to 0} \frac{\gamma(h) - \gamma(0)}{h} = \lim_{h \to 0} \frac{f(a_1 + h, a_2) - f(a_1, a_2)}{h}$$

Compare this with the definition of a single variable derivative

$$\lim_{h \to 0} \frac{f(a + h) - f(a)}{h}$$

We see that when computing a partial derivative, only one of the variables actually changes; the rest stay constant. So to calculate a partial derivative with respect to a particular variable, we just pretend all the other variables are constants and differentiate like normal.

Example. Find the partial derivatives of the function $f(x, y) = x^2 y^3 + 5e^{xy}$. Again, to find the partial derivative with respect to x, just pretend y is a constant then differentiate like it's a single variable function.

$$\frac{\partial f}{\partial x} = 2xy^3 + 5ye^{xy}$$

To find the partial derivative with respect to y, just do the opposite

$$\frac{\partial f}{\partial y} = 3x^2 y^2 + 5xe^{xy}$$

Remember that partial derivatives are just special cases of directional derivatives, so everything we talked about earlier still applies here. For example, since partial derivatives are unit directional derivatives, they can represent the slope of a tangent line on the graph of a real valued function $f(x, y)$. The partial derivative with respect to x at a point $\mathbf{a}$ would be the slope of the tangent line at $f(\mathbf{a})$ which points in the x direction.

Second Partials

Because a partial derivative of a function is itself a function, we can take the partial derivative of a partial derivative. Of course, partial derivatives are just directional derivatives, so more generally we can take the directional derivative of a directional derivative, but those are usually hard to compute.

Anyway, these are called *second order partial derivatives*. Notice that we can take the partial derivative of a function with respect to x twice, with respect to y twice, or with respect to x then with respect to y (or vice versa), so a two variable function has a total of four possible second order partial derivatives.

Example. Find all the second partial derivatives of $f(x, y) = x^5 y^2$.

$$\frac{\partial f}{\partial x} = 5x^4 y^2 \qquad\qquad \frac{\partial f}{\partial y} = 2x^5 y$$

$$\frac{\partial^2 f}{\partial x^2} = 20x^3 y^2 \qquad\qquad \frac{\partial^2 f}{\partial y^2} = 2x^5$$

$$\frac{\partial^2 f}{\partial y \partial x} = 10x^4 y \qquad\qquad \frac{\partial^2 f}{\partial x \partial y} = 10x^4 y$$

Notice that if we take the partial with respect to x then y, or with respect to y then x, the result is the same. These are called *mixed partial derivatives*, and it turns out they are the same pretty much all of the time. To be specific, if the second partial derivatives of a real valued function $f(x, y)$ are continuous at a point $\mathbf{a}$, then the mixed second partials are equal at $\mathbf{a}$. The proof of this statement is too hard to include at this point, but we will have the tools to easily prove it in chapter 6.

Exercises

1. Compute the directional derivative $D_{\mathbf{v}}f(\mathbf{a})$ for the given function, point, and direction vector

   (a) $f(x, y) = 3xy$, $\mathbf{a} = \langle 1, 2 \rangle$, $\mathbf{v} = \langle 3, 5 \rangle$

   (b) $f(x, y) = \langle 2x^2, x + 3y \rangle$, $\mathbf{a} = \langle 1, 1 \rangle$, $\mathbf{v} = \langle -2, 4 \rangle$

2. A duck is walking around on the graph of the function $f(x, y, z) = 2x^2 + yz$. Suppose he is standing at the point on the graph whose corresponding input is $(1, 0, 2)$ and he starts moving in the direction given by $\mathbf{v} = \langle 1, 2, 0 \rangle$. What is the initial slope of his path?

3. Find all first and second partial derivatives of $f(x, y) = 2x^3 y^4 + e^{2xy}$

Answers

1. (a) 33

   (b) $\langle -4, 10 \rangle$

2. $\frac{8}{\sqrt{5}}$

3.

$$\frac{\partial f}{\partial x} = 6x^2 y^4 + 2y e^{2xy} \qquad\qquad \frac{\partial f}{\partial y} = 8x^3 y^3 + 2x e^{2xy}$$

$$\frac{\partial^2 f}{\partial x^2} = 12xy^4 + 4y^2 e^{2xy} \qquad\qquad \frac{\partial^2 f}{\partial y^2} = 24x^3 y^2 + 4x^2 e^{2xy}$$

$$\frac{\partial^2 f}{\partial x \partial y} = 24x^2 y^3 + 2e^{2xy} + 4xy e^{2xy}$$

3.3 Differentials

With directional derivatives and partial derivatives, we can find out how a function is changing around a point in a *certain direction*. We are restricted to changes in a certain direction because when taking a directional derivative we must take it with respect to a specific vector **v**, which represents the direction of movement in the domain. In this section we will see that if a function is differentiable at a point **a**, then there exists a linear approximation to the function around $f(\mathbf{a})$. This linear approximation then tells us about the function's behavior in an *entire neighborhood* around $f(\mathbf{a})$, not just in a certain direction.

Before we dive into the details, we should remember what a differential is for a single variable function. If some function $f(x)$ is differentiable at a, then there exists a linear approximation to the function around $f(a)$. This linear approximation takes the form of a tangent line with an equation

$$y - f(a) = f'(a)(x - a)$$

If we denote the approximate change in the function $y - f(a)$ by df_a and the change in x $(x - a)$ with h, then this becomes

$$df_a(h) = f'(a)h$$

and we call this the differential of f at a. So the differential is the best linear approximation to the function around $f(a)$, and it tells us the approximate change in the function when we move h away from a. The derivative $f'(a)$ is the thing we multiply our displacement h by to get the approximate change in f.

Differentiability

Again, if a function is differentiable, we want there to be a best linear approximation to the function at that point. One of the big ideas of calculus is approximating nonlinear things by linear things because in general linear things are nicer. This linear approximation to the function is going to be a linear transformation, so this is where section 2.4 comes into use. A function $f(\mathbf{x})$ is said to be *differentiable* at a point $\mathbf{a}$ if there exists a linear transformation $L(\mathbf{h})$ such that

$$\lim_{\mathbf{h}\to 0} \frac{f(\mathbf{a}+\mathbf{h}) - f(\mathbf{a}) - L(\mathbf{h})}{|\mathbf{h}|} = \mathbf{0}$$

Here $\mathbf{h}$ represents the displacement from our initial point $\mathbf{a}$. Notice that $f(\mathbf{a}+\mathbf{h}) - f(\mathbf{a})$ is the *actual* change in the function as we move from $\mathbf{a}$ to $\mathbf{a}+\mathbf{h}$, while $L(\mathbf{h})$ is the *approximate change*. So essentially what this definition says is that if our function is differentiable, the difference between the actual change and our approximation to the change, or the error in approximation, approaches zero faster than $|\mathbf{h}|$ does (if $|\mathbf{h}|$ approached zero faster, the fraction would get bigger instead of going to zero).

If f is differentiable, then we call the linear transform $L(\mathbf{h})$ the **differential** of f at $\mathbf{a}$, denoted by $df_{\mathbf{a}}(\mathbf{h})$. The matrix which corresponds to this linear transform is called the **derivative** of f at $\mathbf{a}$.

$$df_{\mathbf{a}}(\mathbf{h}) = f'(\mathbf{a})\mathbf{h}$$

So you can see the similarities with the single variable differential. The differential is a linear approximation of the function at $f(\mathbf{a})$. To approximate the change in the function when we move along the vector $\mathbf{h}$ away from $\mathbf{a}$, we multiply the displacement $\mathbf{h}$ by the derivative $f'(\mathbf{a})$. This multiplication, however, is now matrix multiplication because the derivative is a matrix and the displacement vector $\mathbf{h}$ can be thought of as a single column matrix.

Example. To give a simple example of how this definition can be used, we can prove that a linear function is differentiable and its differential is just the function itself, which is obvious if you think about it. If a function is already linear, there is no better linear approximation. So we need to show that plugging in $L(\mathbf{h}) = f(\mathbf{h})$ will satisfy the definition of differentiability.

$$\lim_{\mathbf{h}\to 0} \frac{f(\mathbf{a}+\mathbf{h}) - f(\mathbf{a}) - f(\mathbf{h})}{|\mathbf{h}|} = \lim_{\mathbf{h}\to 0} \frac{f(\mathbf{a}) + f(\mathbf{h}) - f(\mathbf{a}) - f(\mathbf{h})}{|\mathbf{h}|} = \lim_{\mathbf{h}\to 0} \frac{\mathbf{0}}{|\mathbf{h}|} = \mathbf{0}$$

Here we used the properties of linear transforms to split up $f(\mathbf{a}+\mathbf{h})$.

We want a way to know whether or not a function is differentiable without having to refer directly to the definition of differentiability every time. A function is said to be

continuously differentiable at a point **a** if all of its first partial derivatives exist and are continuous at **a**. It turns out that if a function is continuously differentiable at **a**, then it is differentiable at **a**. We will not prove this here, but this condition can be very convenient for determining the differentiability of a function.

Before we can actually start computing some differentials, we will need a way to compute the derivative matrix of a function. And before we learn how to do that, we will need to talk about the relationship between differentials and directional derivatives which we hinted at in the beginning of this section.

Directional Derivatives

Taking the directional derivative of a function with respect to a vector **v** in the domain tells us how the function is changing only in that particular direction. However, in the differential we can get information about how the function changes in *any* direction **h** (**v** and **h** both just represent some displacement out from the point at which we take the derivative **a**).

Notice the difference in notation between the two operations: the directional derivative $D_{\mathbf{v}}f(\mathbf{a})$ is a function of **a**. So we must choose a direction and we are restricted to that direction, but we can choose any point **a** along the line in the domain determined by **v**. The differential $df_{\mathbf{a}}(\mathbf{h})$, on the other hand, is a function of **h**. So we must choose a point and we are restricted to the neighborhood around that point **a**, but we can look at the function's behavior in any direction **h** out from **a** that we want.

Since both of these things really just tell us how a function is changing in some direction out from **a**, you might expect $D_{\mathbf{v}}f(\mathbf{a})$ to have the same value as $df_{\mathbf{a}}(\mathbf{v})$ as long as f is differentiable. This makes sense because they're both supposed to approximate the change in f when we walk along the vector **v** out from **a**. It turns out that this is always true.

Theorem. If f is differentiable at **a**, then all possible directional derivatives at **a** exist and

$$D_{\mathbf{v}}f(\mathbf{a}) = df_{\mathbf{a}}(\mathbf{v})$$

Proof. We start with the definition of differentiability, using the differential in place of $L(\mathbf{h})$ because we know the function is differentiable.

$$\lim_{\mathbf{h}\to 0} \frac{f(\mathbf{a}+\mathbf{h}) - f(\mathbf{a}) - df_{\mathbf{a}}(\mathbf{h})}{|\mathbf{h}|} = \mathbf{0}$$

Now we make the substitution $\mathbf{h} = t\mathbf{v}$. We do this because the definition of a directional derivative involves a limit as a scalar goes to 0, not a vector. Note that if **h** approaches

0, we can also say that t approaches 0.

$$\lim_{t \to 0} \frac{f(\mathbf{a} + t\mathbf{v}) - f(\mathbf{a}) - df_{\mathbf{a}}(t\mathbf{v})}{|t\mathbf{v}|} = \frac{1}{|\mathbf{v}|} \lim_{t \to 0} \left[\frac{f(\mathbf{a} + t\mathbf{v}) - f(\mathbf{a})}{t} - \frac{df_{\mathbf{a}}(t\mathbf{v})}{t} \right] = \mathbf{0}$$

In the second step we moved the constant $1/|\mathbf{v}|$ out of the limit (which we can then get rid of because the other side is zero) and split the fraction. We can now use the fact that the differential is a linear transform to move t out of the differential, making it cancel out with the t in the denominator.

$$\lim_{t \to 0} \left[\frac{f(\mathbf{a} + t\mathbf{v}) - f(\mathbf{a})}{t} \right] - df_{\mathbf{a}}(\mathbf{v}) = \mathbf{0}$$

$$df_{\mathbf{a}}(\mathbf{v}) = \lim_{t \to 0} \frac{f(\mathbf{a} + t\mathbf{v}) - f(\mathbf{a})}{t} = D_{\mathbf{v}} f(\mathbf{a})$$

Where we have recognized that big limit as the definition of a directional derivative. $\square$

This fact not only makes it very easy to compute directional derivatives (as we will see next section), but also gives us a way to check if a function is differentiable. If we happen to find a vector $\mathbf{v}$ such that

$$D_{\mathbf{v}} f(\mathbf{a}) \neq df_{\mathbf{a}}(\mathbf{v})$$

then that proves the function is *not* differentiable. However, note that we can't use directional derivatives to prove that a function *is* differentiable. A function can have all of its directional derivatives at a point $\mathbf{a}$ yet still not be differentiable there.

Derivatives

Remember from section 2.4 that the columns of the matrix of a linear transform are determined by plugging the standard basis vectors into the linear transform. So to get the first column of our derivative matrix, we should compute $df_{\mathbf{a}}(\mathbf{e}_1)$. By the last theorem, for a differentiable function this is the same thing as $D_{\mathbf{e}_1} f(\mathbf{a})$. But this is just the partial derivative of f with respect to x!. So the first column of the derivative matrix is just the partial derivatives of all the coordinate functions with respect to x. The second column will be the partial derivatives of the coordinate functions with respect to y, and so on. For a function $f : \mathbb{R}^3 \mapsto \mathbb{R}^3$ given by $f(x, y, z) = \langle f_1, f_2, f_3 \rangle$, the 3×3 derivative matrix at $\mathbf{a}$ looks like this:

$$f'(\mathbf{a}) = \begin{bmatrix} \dfrac{\partial f_1}{\partial x} & \dfrac{\partial f_1}{\partial y} & \dfrac{\partial f_1}{\partial z} \\[2ex] \dfrac{\partial f_2}{\partial x} & \dfrac{\partial f_2}{\partial y} & \dfrac{\partial f_2}{\partial z} \\[2ex] \dfrac{\partial f_3}{\partial x} & \dfrac{\partial f_3}{\partial y} & \dfrac{\partial f_3}{\partial z} \end{bmatrix}$$

where all of the partial derivatives are evaluated at $\mathbf{a}$. So you can see that the derivative matrix is made up of first order partial derivatives of the function. Each separate coordinate function gets its own row, and each independent variable gets its own column. So for a general multivariable vector function $f : \mathbb{R}^n \mapsto \mathbb{R}^m$, the derivative matrix, also called the ***Jacobian***, will have dimensions $m \times n$: m rows for the m output functions, and n columns for the n independent variables.

Example. The Jacobian of the function $f(x, y) = \langle 3x^2 + y, 2x + 5y^3 \rangle$ is

$$\begin{bmatrix} 6x & 1 \\ 2 & 15y^2 \end{bmatrix}$$

The first partials are continuous, so this function is continuously differentiable and therefore differentiable. This implies that the Jacobian is indeed the derivative matrix for this function. You can compute the Jacobian for any function, but if it is not differentiable it doesn't necessarily have any meaning.

As a last example, we see how the differential can be used to actually approximate the change in a function.

Example. Consider the function $f(x, y) = x^2 + y^2$. Estimate the value of $f(1.2, 1.3)$. To start out, we compute the differential at $\mathbf{a} = (1, 1)$.

$$f'(1, 1) = \begin{bmatrix} 2x & 2y \end{bmatrix} = \begin{bmatrix} 2 & 2 \end{bmatrix}$$

$$df_{(1,1)}(\mathbf{h}) = f'(1, 1)\mathbf{h} = 2h_1 + 2h_2$$

The displacement vector out from $(1, 1)$ we are interested in is $\mathbf{h} = \langle .2, .3 \rangle$. So the approximate change in the function going from $(1, 1)$ to $(1.2, 1.3)$ is

$$f(1.2, 1.3) - f(1, 1) \approx 2(.2) + 2(.3) = 1$$

$$f(1.2, 1.3) \approx 1 + f(1, 1) = 3$$

This is fairly close to the actual value of 3.13

Exercises

1. Using the definition of differentiability, show that if a function $f(\mathbf{x})$ is constant, then it is differentiable everywhere and its differential is zero

2. Consider the function defined by

$$f(x, y) = \begin{cases} \dfrac{x^2 y}{x^2 + y^2} & \text{if } (x, y) \neq (0, 0) \\ 0 & \text{if } (x, y) = (0, 0) \end{cases}$$

(a) Show that the function is continuous at the origin

(b) Compute the partial derivatives at $(0,0)$ directly from the definition of a directional derivative

(c) Compute the directional derivative at $(0,0)$ with respect to $\mathbf{v} = \langle 1, 1 \rangle$

(d) Show that all directional derivatives exist at the origin by finding a formula in terms of an arbitrary vector $\mathbf{v} = \langle a, b \rangle$

(e) Conclude that the function is still not differentiable, despite all the directional derivatives existing.

3. Compute the Jacobian

(a) $f(x, y) = \langle 3x^4 y^2, \sin(x^2), e^{xy^2} \rangle$

(b) $f(x, y, z) = 5x^3 z + y^2 z^4$

4. Use a differential to approximate $(3.04)^2 + (0.98)^2 + (4.97)^2$

Answers

1. If f is a constant vector function, then for all $\mathbf{x}$ it equals some constant vector $\mathbf{b}$. Using $L(\mathbf{h}) = \mathbf{0}$,

$$\lim_{\mathbf{h} \to 0} \frac{f(\mathbf{a} + \mathbf{h}) - f(\mathbf{a}) - \mathbf{0}}{|\mathbf{h}|} = \lim_{\mathbf{h} \to 0} \frac{\mathbf{b} - \mathbf{b}}{|\mathbf{h}|} = \mathbf{0}$$

2. (a) Use the polar substitution trick to convert it to a single variable limit

$$\lim_{r \to 0} \frac{r^3 \cos \theta \sin \theta}{r^2} = 0 = f(0,0)$$

(b) $D_1 f(0,0) = D_2 f(0,0) = 0$

(c) $\frac{1}{2}$

(d) $D_{\mathbf{v}} f(0,0) = \dfrac{a^2 b}{a^2 + b^2}$

(e) Since the partial derivatives are both 0, the differential is 0 for all displacements, i.e.

$$df_{(0,0)}(\mathbf{h}) = 0$$

for all $\mathbf{h}$. But from parts (c) and (d) we can see that there are directional derivatives at the origin with nonzero values, so the function cannot be differentiable

3. (a)

$$\begin{bmatrix} 12x^3y^2 & 6x^4y \\ 2x\cos(x^2) & 0 \\ y^2e^{xy^2} & 2xye^{xy^2} \end{bmatrix}$$

(b)

$$\begin{bmatrix} 15x^2z & 2yz^4 & 5x^3 + 4y^2z^3 \end{bmatrix}$$

4. Use the function $f(x, y, z) = x^2 + y^2 + z^2$ with $\mathbf{a} = \langle 3, 1, 5 \rangle$ and $\mathbf{h} = \langle .04, -.02, -.03 \rangle$; the result is

$$df_\mathbf{a}(\mathbf{h}) = -.1$$

$$f(\mathbf{a} + \mathbf{h}) \approx f(\mathbf{a}) - .1 = 34.9$$

3.4 Gradients

In this section we will look at a very important vector which we will use often throughout the rest of the book. To introduce the concept, consider the derivative matrix of any real valued function $f(\mathbf{x})$. If this function is $f : \mathbb{R}^n \mapsto \mathbb{R}$, then its Jacobian will have dimensions $1 \times n$. In other words, it will be a row vector of the first partial derivatives of f. We call this vector the **gradient** of the function.

$$\nabla f(\mathbf{x}) = \langle \frac{\partial f}{\partial x_1}, \frac{\partial f}{\partial x_2}, ..., \frac{\partial f}{\partial x_n} \rangle$$

So when we take the gradient of a *real valued function*, it gives us a *vector* consisting of the first partials of f. Again, this is really just the derivative matrix of a real valued function represented as a vector. Note that this also implies that we can only take the gradient vector of a differentiable function.

Example. Find the gradient vector of $f(x, y, z) = 2x^2y + 5y^2z$

$$\nabla f(x, y, z) = \langle 4xy, 2x^2 + 10yz, 5y^2 \rangle$$

Gradients and Directional Derivatives

Calculating directional derivatives becomes a lot simpler with this gradient vector. In the last section, we learned that for a *differentiable function*, the directional derivative at $\mathbf{a}$ with respect to $\mathbf{v}$ is the same as the differential at $\mathbf{a}$ evaluated at $\mathbf{v}$.

$$D_\mathbf{v}f(\mathbf{a}) = df_\mathbf{a}(\mathbf{v})$$

For a general three variable real valued function, the differential will look like

$$df_{\mathbf{a}}(\mathbf{v}) = \begin{bmatrix} \dfrac{\partial f}{\partial x} & \dfrac{\partial f}{\partial y} & \dfrac{\partial f}{\partial z} \end{bmatrix} \begin{bmatrix} v_1 \\ v_2 \\ v_3 \end{bmatrix}$$

$$= \frac{\partial f}{\partial x} v_1 + \frac{\partial f}{\partial y} v_2 + \frac{\partial f}{\partial z} v_3$$

$$= \nabla f(\mathbf{a}) \cdot \mathbf{v}$$

Recall that matrix multiplication between a row vector and a column vector can be represented as a dot product between them. Therefore, we can compute the differential of a *real valued function* by taking the dot product of the gradient vector, which is really the derivative, and the displacement vector. Therefore, for a *differentiable real valued function*, we can compute *directional derivatives* as such:

$$D_{\mathbf{v}}f(\mathbf{a}) = \nabla f(\mathbf{a}) \cdot \mathbf{v}$$

In the case of a vector valued function, the directional derivatives of the coordinate functions can be computed individually in this manner and then combined to get the overall directional derivative (since directional derivatives can be computed coordinatewise).

Example. Find the directional derivative with respect to $\mathbf{v} = \langle 3, 1 \rangle$ at $\mathbf{a} = (1, 2)$ of $f(x, y) = \langle 3x^3 y, 2xy^4 \rangle$. First we will find the directional derivatives of the coordinate functions f_1 and f_2 individually.

$$\nabla f_1(1, 2) = \langle 9x^2 y, 3x^3 \rangle$$

$$= \langle 18, 3 \rangle$$

So the directional derivative of f_1 at $(1, 2)$ with respect to $\mathbf{v}$ is

$$D_{\mathbf{v}}f_1(1, 2) = \nabla f_1(1, 2) \cdot \mathbf{v} = 18(3) + 3(1) = 57$$

We repeat the process for f_2

$$\nabla f_2(1, 2) = \langle 2y^4, 8xy^3 \rangle$$

$$= \langle 32, 64 \rangle$$

$$D_{\mathbf{v}}f_2(1, 2) = \nabla f_2(1, 2) \cdot \mathbf{v} = 32(3) + 64(1) = 160$$

Now the directional derivative of the overall vector function is

$$D_{\mathbf{v}}f(1, 2) = \langle D_{\mathbf{v}}f_1(1, 2), D_{\mathbf{v}}f_2(1, 2) \rangle$$

$$= \langle 57, 160 \rangle$$

Rate of Change

Expressing directional derivatives in this way also reveals another important property of the gradient vector. Say we are at the point **a** and we want to know: in which direction does the function increase the fastest? For example, suppose the function $H(x, y)$ gives the elevation at a point on a mountain (x, y), and we are standing at the point **a**. We want to know in which direction to walk in order to climb the fastest, or in other words the direction with the most rapid increase in elevation.

Since we are only concerned with what *direction* to move in, we need to look at the unit directional derivative. In particular, for what direction **u** does the unit directional derivative at the point **a** achieve a maximum? Remember that we can write a dot product in terms of the angle between the two vectors.

$$D_{\mathbf{u}}f(\mathbf{a}) = \nabla f(\mathbf{a}) \cdot \mathbf{u} = |\nabla f(\mathbf{a})| \, |\mathbf{u}| \cos\theta = |\nabla f(\mathbf{a})| \cos\theta$$

where θ is the angle between the gradient vector and **u**. We see that the directional derivative achieves a *maximum* when $\theta = 0$. This implies that the gradient and **u** are parallel, so to maximize the rate of change of the function we should choose **u** to point in the same direction as the gradient, i.e.

$$\mathbf{u} = \frac{\nabla f(\mathbf{a})}{|\nabla f(\mathbf{a})|}$$

In other words, the gradient vector represents the direction of fastest increase of the function at the point **a**. Moreover, the value of this rate of change will be $|\nabla f(\mathbf{a})|$ as long as we are taking the directional derivative with respect to a *unit vector*.

A similar argument applies for the direction of fastest *decrease*: the directional derivative achieves a minimum when $\theta = \pi$. Therefore, to get the most rapid decrease of the function, we should choose **u** to point directly opposite of the gradient.

Notice also the significance if the gradient at a point is the zero vector.

$$\nabla f(\mathbf{a}) = \mathbf{0}$$

This means that at **a** there is *no* direction of fastest increase or fastest decrease. We call such a point **a** where the gradient is the zero vector a **critical point**. Critical points are good candidates for places where a function achieves a *local* maximum or minimum, since at these peaks or valleys there will be no direction of fastest increase or decrease. However, critical points could also be neither maxima nor minima, just as in single variable calculus. We will learn later in the chapter how to determine what a critical point represents.

Exercises

1. Find the gradient vector of $f(x,y) = 3xy^3 + e^{x^2 y}$

2. Calculate the directional derivative at $\mathbf{a} = \langle 1, 1, 2 \rangle$ with respect to $\mathbf{v} = \langle 3, 2, 5 \rangle$ of the function

$$f(x, y, z) = \langle 4x^2 yz, xy^3 + 5z \rangle$$

3. A bear is looking for some berries to eat. The function

$$B(x, y) = 2xy + e^{x+y}$$

gives the concentration of berries at the location (x, y). If the bear is currently standing at $(-2, 2)$, in what direction should he move to get the most berries?

    (a) Could there be a place with the maximum concentration of berries?

Answers

1. $\nabla f(x, y) = \langle 3y^3 + 2xye^{x^2 y}, 9xy^2 + x^2 e^{x^2 y} \rangle$

2. $\langle 84, 34 \rangle$

3. $\langle \frac{5}{\sqrt{34}}, -\frac{3}{\sqrt{34}} \rangle$

    (a) No, the gradient vector is never the zero vector

3.5   The Chain Rule

We would like to have a method for computing the differentials and derivatives of compositions of functions, similar to the single variable chain rule. Recall that this rule says

$$\frac{d}{dx}[f(g(x))] = f'(g(x))g'(x)$$

To get the derivative of the composition, we simply multiply the derivatives of the two functions involved. However, note the point at which we evaluate the derivatives: the derivative of f is evaluated at the point $g(x)$, not simply x. The multivariable chain rule will be very similar to this, except that since the derivatives are now matrices it will involve matrix multiplication.

Suppose we have two differentiable functions f and g, where the dimensions of the output of g match up with the dimensions of the input of f. So if g produces a vector with three components, then f should take in a vector with three components. Then we can take the composition of the functions $f \circ g$, which we will denote by C. So $C = f(g(\mathbf{x}))$ is another function, and our goal is to find the differential and derivative of it.

If the input moves from the point $\mathbf{a}$ to the point $\mathbf{a} + \mathbf{h}$, what effect will this have on the entire function C?

$$\Delta C = f(g(\mathbf{a} + \mathbf{h})) - f(g(\mathbf{a}))$$

This change in the input causes a chain reaction to go off: it first directly causes a change in the function g, then that goes on to cause a change in the function f. Since g is differentiable, we can approximate its response to this input change with its differential.

$$g(\mathbf{a} + \mathbf{h}) - g(\mathbf{a}) \approx dg_{\mathbf{a}}(\mathbf{h})$$

$$g(\mathbf{a} + \mathbf{h}) \approx g(\mathbf{a}) + dg_{\mathbf{a}}(\mathbf{h})$$

Plugging in this new value of g after the change into our original equation, we get

$$\Delta C \approx f(g(\mathbf{a}) + dg_{\mathbf{a}}(\mathbf{h})) - f(g(\mathbf{a}))$$

Now at the second link in the chain, we see that f is changing from $g(\mathbf{a})$ to $g(\mathbf{a}) + dg_{\mathbf{a}}(\mathbf{h})$. We can again approximate f's response to this change with its differential.

$$\Delta C \approx df_{g(\mathbf{a})}(dg_{\mathbf{a}}(\mathbf{h}))$$

This suggests that the differential of the function C is just the *composition* of the differentials of f and g. This is the chain rule for multivariable functions that we wanted.

Theorem. Let f and g be functions where g is differentiable at $\mathbf{a}$ and f is differentiable at $g(\mathbf{a})$. Then their composition $C = f \circ g$ is differentiable at $\mathbf{a}$, and its differential is the composition of the differentials of f and g

$$dC_{\mathbf{a}}(\mathbf{h}) = (df_{g(\mathbf{a})} \circ dg_{\mathbf{a}})(\mathbf{h})$$

The derivative of C, the corresponding matrix to its differential, is the matrix product of the derivatives of f and g

$$C'(\mathbf{a}) = f'(g(\mathbf{a}))g'(\mathbf{a})$$

The derivative part follows from the fact that the matrix of a composition of linear transforms is just the product of the matrices of the two transforms involved, which we saw in section 2.4. Note again which points we evaluate the derivatives at: f' is evaluated at $g(\mathbf{a})$, while g' is evaluated at $\mathbf{a}$.

This multivaraible chain rule seems very simple at first glance: the differential of the composition is the composition of the differentials, and the derivative of the composition is the product of the derivatives. However, it can often be tricky to apply, so we will go through some examples.

Example. Suppose we have the functions $f : \mathbb{R}^2 \mapsto \mathbb{R}$, $f(x,y)$, and $g : \mathbb{R} \mapsto \mathbb{R}^2$, $g(t) = \langle g_1, g_2 \rangle$. Then the composition function will be a real valued single variable function $C : \mathbb{R} \mapsto \mathbb{R}$, $C(t)$. To compute its derivative, first compute the Jacobians of f and g.

$$f'(x,y) = \begin{bmatrix} \dfrac{\partial f}{\partial x} & \dfrac{\partial f}{\partial y} \end{bmatrix} \qquad g'(t) = \begin{bmatrix} \dfrac{dg_1}{dt} \\ \dfrac{dg_2}{dt} \end{bmatrix}$$

The derivative of C, which is just a 1×1 matrix (a real number), will be their product

$$C'(t) = f'(g(t))g'(t)$$

$$\begin{bmatrix} \dfrac{dC}{dt} \end{bmatrix} = \begin{bmatrix} \dfrac{\partial f}{\partial x} & \dfrac{\partial f}{\partial y} \end{bmatrix} \begin{bmatrix} \dfrac{dg_1}{dt} \\ \dfrac{dg_2}{dt} \end{bmatrix}$$

Again, matrix multiplication between a row vector and column vector can be represented by the dot product:

$$\frac{dC}{dt} = \nabla f(g(t)) \cdot g'(t)$$

If we write it out, we get this expression:

$$\frac{dC}{dt} = \frac{\partial f}{\partial x}\frac{dg_1}{dt} + \frac{\partial f}{\partial y}\frac{dg_2}{dt}$$

We must remember that the partial derivatives of f are evaluated at $g(t)$, something which the notation can fail to convey.

Example. Use the results of the previous example to calculate $\frac{dC}{dt}$ where C is the composition of the functions $f(x,y) = xy^2$ and $g(t) = \langle 2t^2, 3t \rangle$.

$$\begin{aligned} \frac{dC}{dt} &= \frac{\partial f}{\partial x}\frac{dg_1}{dt} + \frac{\partial f}{\partial y}\frac{dg_2}{dt} \\ &= (y^2)(4t) + (2xy)(3) \\ &= (9t^2)(4t) + (12t^3)(3) \\ &= 36t^3 + 36t^3 = 72t^3 \end{aligned}$$

Example. Suppose we have the functions $f : \mathbb{R}^3 \mapsto \mathbb{R}$, $f(x,y,z)$, and $g : \mathbb{R}^2 \mapsto \mathbb{R}^3$, $g(u,v) = \langle g_1, g_2, g_3 \rangle$. Then the composition of the functions will be a two variable real valued function $C : \mathbb{R}^2 \mapsto \mathbb{R}$, $C(u,v)$. We use the chain rule to compute the partial derivatives of this function.

$$f'(x, y, z) = \begin{bmatrix} \dfrac{\partial f}{\partial x} & \dfrac{\partial f}{\partial y} & \dfrac{\partial f}{\partial z} \end{bmatrix} \qquad g'(u, v) = \begin{bmatrix} \dfrac{\partial g_1}{\partial u} & \dfrac{\partial g_1}{\partial v} \\[2mm] \dfrac{\partial g_2}{\partial u} & \dfrac{\partial g_2}{\partial v} \\[2mm] \dfrac{\partial g_3}{\partial u} & \dfrac{\partial g_3}{\partial v} \end{bmatrix}$$

The derivative of C will therefore be a 1×2 matrix

$$C'(u, v) = \begin{bmatrix} \dfrac{\partial C}{\partial u} & \dfrac{\partial C}{\partial v} \end{bmatrix} = f'(g(u, v))g'(u, v)$$

If we write out the expressions we get this:

$$\frac{\partial C}{\partial u} = \frac{\partial f}{\partial x}\frac{\partial g_1}{\partial u} + \frac{\partial f}{\partial y}\frac{\partial g_2}{\partial u} + \frac{\partial f}{\partial z}\frac{\partial g_3}{\partial u}$$

$$\frac{\partial C}{\partial v} = \frac{\partial f}{\partial x}\frac{\partial g_1}{\partial v} + \frac{\partial f}{\partial y}\frac{\partial g_2}{\partial v} + \frac{\partial f}{\partial z}\frac{\partial g_3}{\partial v}$$

Example. Use the above example to calculate the partial derivatives of $C(u, v)$ where C is the composition of the functions $f(x, y, z) = xyz$ and $g(u, v) = \langle 3u + v, 2u^2, v^3 \rangle$.

$$\frac{\partial C}{\partial u} = \frac{\partial f}{\partial x}\frac{\partial g_1}{\partial u} + \frac{\partial f}{\partial y}\frac{\partial g_2}{\partial u} + \frac{\partial f}{\partial z}\frac{\partial g_3}{\partial u}$$
$$= (yz)(3) + (xz)(4u) + (xy)(0)$$
$$= (2u^2)(v^3)(3) + (3u + v)(v^3)(4u)$$
$$= 6u^2v^3 + 12u^2v^3 + 4uv^4 = 18u^2v^3 + 4uv^4$$

$$\frac{\partial C}{\partial v} = \frac{\partial f}{\partial x}\frac{\partial g_1}{\partial v} + \frac{\partial f}{\partial y}\frac{\partial g_2}{\partial v} + \frac{\partial f}{\partial z}\frac{\partial g_3}{\partial v}$$
$$= (yz)(1) + (xz)(0) + (xy)(3v^2)$$
$$= (2u^2)(v^3)(1) + (3u + v)(2u^2)(3v^2)$$
$$= 2u^2v^3 + 18u^3v^2 + 6u^2v^3 = 8u^2v^3 + 18u^3v^2$$

Double Chain Rule

Now we will consider how to find second partials of a composition of functions using the chain rule; this is when things start to really get messy. The idea behind this method is that the partial derivative of a function is yet another function. Therefore, if it is differentiable, then it has its own derivative matrix.

To introduce this idea, consider the partial derivative of some random two variable, real valued function $\frac{\partial f}{\partial x}$. If this is itself a differentiable function, what is its derivative matrix? Since the original function is $f : \mathbb{R}^2 \mapsto \mathbb{R}$, we know that this partial derivative will also be a function $\frac{\partial f}{\partial x} : \mathbb{R}^2 \mapsto \mathbb{R}$. So its derivative matrix will also have dimensions 1×2. As with any other derivative matrix, to get the element in the first column we take the partial derivative with respect to the first independent variable, in this case x.

$$\frac{\partial}{\partial x}\left(\frac{\partial f}{\partial x}\right) = \frac{\partial^2 f}{\partial x^2}$$

The element in the second column comes from partially differentiating with respect to y

$$\frac{\partial}{\partial y}\left(\frac{\partial f}{\partial x}\right) = \frac{\partial^2 f}{\partial y \partial x}$$

Therefore, the derivative matrix of the first order partial derivative $\frac{\partial f}{\partial x}$ will be

$$\frac{\partial f}{\partial x}'(x, y) = \begin{bmatrix} \frac{\partial^2 f}{\partial x^2} & \frac{\partial^2 f}{\partial y \partial x} \end{bmatrix}$$

Example. Let $C(r, \theta)$ be the composition of the functions $f(x, y)$ and $g(r, \theta) = \langle r\cos\theta, r\sin\theta \rangle$. Find $\frac{\partial^2 C}{\partial \theta \partial r}$. First we find the first partial derivative of C with respect to r.

$$C'(r, \theta) = f'(g(r, \theta))g'(r, \theta)$$

$$\begin{bmatrix} \frac{\partial C}{\partial r} & \frac{\partial C}{\partial \theta} \end{bmatrix} = \begin{bmatrix} \frac{\partial f}{\partial x} & \frac{\partial f}{\partial y} \end{bmatrix} \begin{bmatrix} \cos\theta & -r\sin\theta \\ \sin\theta & r\cos\theta \end{bmatrix}$$

$$\frac{\partial C}{\partial r} = \frac{\partial f}{\partial x}\cos\theta + \frac{\partial f}{\partial y}\sin\theta$$

To get the second partial that we want, we need to differentiate both sides of this expression with respect to θ.

$$\frac{\partial^2 C}{\partial \theta \partial r} = \frac{\partial}{\partial \theta}\left(\frac{\partial f}{\partial x}\cos\theta\right) + \frac{\partial}{\partial \theta}\left(\frac{\partial f}{\partial y}\sin\theta\right)$$

For both terms on the right side, we use the product rule

$$\frac{\partial^2 C}{\partial \theta \partial r} = \frac{\partial}{\partial \theta}\left(\frac{\partial f}{\partial x}\right)\cos\theta - \frac{\partial f}{\partial x}\sin\theta + \frac{\partial}{\partial \theta}\left(\frac{\partial f}{\partial y}\right)\sin\theta + \frac{\partial f}{\partial y}\cos\theta$$

To differentiate $\frac{\partial f}{\partial x}$ and $\frac{\partial f}{\partial y}$ with respect to θ, we need to use the chain rule again. First, let's focus on $\frac{\partial}{\partial \theta}\frac{\partial f}{\partial x}$. We are really trying to differentiate the composition $\frac{\partial f}{\partial x} \circ g(r, \theta)$ with respect to θ, so let's write out the chain rule for this

$$\begin{bmatrix} \frac{\partial}{\partial r}\frac{\partial f}{\partial x} & \frac{\partial}{\partial \theta}\frac{\partial f}{\partial x} \end{bmatrix} = \begin{bmatrix} \frac{\partial^2 f}{\partial x^2} & \frac{\partial^2 f}{\partial y \partial x} \end{bmatrix} \begin{bmatrix} \cos\theta & -r\sin\theta \\ \sin\theta & r\cos\theta \end{bmatrix}$$

Therefore, we see that

$$\frac{\partial}{\partial\theta}\frac{\partial f}{\partial x} = -\frac{\partial^2 f}{\partial x^2}r\sin\theta + \frac{\partial^2 f}{\partial y\partial x}r\cos\theta$$

Similarly, to find $\frac{\partial}{\partial\theta}\frac{\partial f}{\partial y}$, we are really trying to find the partial derivative of the composition $\frac{\partial f}{\partial y} \circ g(r,\theta)$ with respect to θ.

$$\begin{bmatrix} \dfrac{\partial}{\partial r}\dfrac{\partial f}{\partial y} & \dfrac{\partial}{\partial\theta}\dfrac{\partial f}{\partial y} \end{bmatrix} = \begin{bmatrix} \dfrac{\partial^2 f}{\partial x\partial y} & \dfrac{\partial^2 f}{\partial y^2} \end{bmatrix}\begin{bmatrix} \cos\theta & -r\sin\theta \\ \sin\theta & r\cos\theta \end{bmatrix}$$

$$\frac{\partial}{\partial\theta}\frac{\partial f}{\partial y} = -\frac{\partial^2 f}{\partial x\partial y}r\sin\theta + \frac{\partial^2 f}{\partial y^2}r\cos\theta$$

Now all that's left is to plug these into the expression we had, then simplify, remembering that the mixed partials will be equal.

$$\frac{\partial^2 C}{\partial\theta\partial r} = \left(-\frac{\partial^2 f}{\partial x^2}r\sin\theta + \frac{\partial^2 f}{\partial y\partial x}r\cos\theta\right)\cos\theta - \frac{\partial f}{\partial x}\sin\theta +$$

$$\left(-\frac{\partial^2 f}{\partial x\partial y}r\sin\theta + \frac{\partial^2 f}{\partial y^2}r\cos\theta\right)\sin\theta + \frac{\partial f}{\partial y}\cos\theta$$

$$= r\cos\theta\sin\theta\left(\frac{\partial^2 f}{\partial y^2} - \frac{\partial^2 f}{\partial x^2}\right) + r(\cos^2\theta - \sin^2\theta)\frac{\partial^2 f}{\partial x\partial y} + \frac{\partial f}{\partial y}\cos\theta - \frac{\partial f}{\partial x}\sin\theta$$

Exercises

1. Find the derivative of the composition $f \circ g$ if

$$f(x,y) = x^2 + y^2 \qquad \text{and} \qquad g(t) = \langle 3t^3 + 2, 5t\rangle$$

2. Find the partial derivatives of the composition $f \circ g$ if

$$f(x,y) = x^2 + y^2 \qquad \text{and} \qquad g(u,v) = \langle 2uv, 3u^2 + v\rangle$$

3. Show that if $f(x,y)$ is a real valued function, then

$$\nabla(f^n) = nf^{n-1}\nabla f$$

Answers

1. $\frac{dC}{dt} = 54t^5 + 36t^2 + 50t$

2. $\frac{\partial C}{\partial u} = 8uv^2 + 36u^3 + 12uv$ and $\frac{\partial C}{\partial v} = 8u^2 v + 6u^2 + 2v$

3. Let $C = s \circ f$ denote the composition of the functions $s(u) = u^n$ and $f(x,y)$. The gradient of the composition is just its derivative matrix.

$$\begin{bmatrix} \dfrac{\partial C}{\partial x} & \dfrac{\partial C}{\partial y} \end{bmatrix} = \begin{bmatrix} nu^{n-1} \end{bmatrix} \begin{bmatrix} \dfrac{\partial f}{\partial x} & \dfrac{\partial f}{\partial y} \end{bmatrix}$$
$$= \begin{bmatrix} nf^{n-1}\dfrac{\partial f}{\partial x} & nf^{n-1}\dfrac{\partial f}{\partial y} \end{bmatrix}$$
$$= nf^{n-1}\langle \dfrac{\partial f}{\partial x}, \dfrac{\partial f}{\partial y} \rangle = nf^{n-1}\nabla f$$

3.6   The Implicit Function Theorem

In single variable calculus, you learned about implicit differentiation, which you used to find the derivative of things such as the equation of a circle $x^2+y^2 = 1$. You differentiated with respect to x, making sure to stick on a $\frac{dy}{dx}$ after differentiating any term involving y. You could then solve for the derivative, which gave you the slope of the tangent line to the circle at some point.

You should notice something wrong with this: the derivative only gives us the slope of the tangent line when we are talking about the *graph* of some *function*. Clearly, the circle is not the graph of a function because it fails the vertical line test. Rather, it can be thought of as a level set of the function $G(x,y) = x^2+y^2-1$ where G is set to equal zero, which we will call the *zero set* from now on. So how can we talk about the derivative of the circle? What even is implicit differentiation?!

Suppose we wanted to find the slope of the tangent line to the circle at some point **a** which lies somewhere in the top half of the circle. If we are only interested in finding the slope of the tangent line at this point **a**, do we really care what's happening at the rest of the circle? No! All we care about is the little section of the circle immediately surrounding **a**.

So maybe we can find a new *function* whose graph exactly matches the circle around the point **a**, but not necessarily around the whole circle. If we can find such a function, then it does make sense to take the derivative of it, and the derivative at **a** will give us the slope of the tangent line to the graph of the function and therefore the circle. For a circle, this function is very easy to find. By simply solving for y in the equation of the unit circle, we find that

$$f(x) = \sqrt{1-x^2}$$

is a function whose graph is exactly the upper half of the circle (but not the lower half). We might say that this *implicit function* is *implied* by the equation of the circle $x^2+y^2 = 1$.

This is the big idea of the ***implicit function theorem***. Given a curve or surface which is the zero set of some function, it might be impossible to represent the whole set as the graph of a function. However, if we are only interested in a little part of the set around some point $\mathbf{a}$, then we can find a function whose graph exactly matches the set *only around that point*. In the case of a circle, we easily found an expression for that implicit function, but the theorem guarantees that these functions exist even if we cannot find a direct expression for them.

Theorem. Suppose we have a curve or surface which can be represented as the zero set of some real valued function $G(\mathbf{x}) = 0$. If the ith partial derivative of G is *nonzero* at a point $\mathbf{a}$, then we can solve for the ith independent variable as a differentiable function of the other independent variables. The graph of this new implicit function will match the curve/surface in a local neighborhood around $\mathbf{a}$.

The meaning of this will become clearer when we go through some examples, so let's go back to the circle $x^2 + y^2 = 1$. This curve can be represented as the zero set of $G(x, y) = x^2 + y^2 - 1$. We want to find a function whose graph matches the circle around some point $\mathbf{a}$, say $(0, 1)$. If we are trying to solve for y as a function of x, then the theorem says that the second partial derivative of G must be nonzero at $(0, 1)$. But

$$\frac{\partial G}{\partial y} = 2y = 2$$

so this condition is satisfied. Therefore, a differentiable function $f(x)$ exists such that the graph of this function matches the circle in some neighborhood around $(0, 1)$. In other words, the function satisfies $G(x, f(x)) = 0$ around $(0, 1)$.

Of course, the function guaranteed by the implicit function theorem in this case is the same one that we talked about earlier:

$$f(x) = \sqrt{1 - x^2}$$

has a graph which matches the top half of the circle.

To illustrate what happens if the second partial derivative were zero, suppose we wanted to find a function which matches the circle around $(1, 0)$. Then the second partial is zero so we cannot solve for y as a function of x. This makes sense intuitively because a small section of the circle around $(1, 0)$ would fail the vertical line test, and therefore cannot be a function of x.

We can, however, solve for x as a function of y because the *first* partial derivative of G is nonzero at this point. In particular,

$$x = \sqrt{1 - y^2}$$

is a function whose graph matches the right half of the circle. So we see that at every point on the circle there exists some function, either $f(x)$ or $f(y)$, whose graph matches the circle locally.

Implicit Differentiation

Notice that the implicit function theorem also says that the new function will be differentiable. This provides justification for implicit differentiation. We know that at any point on the circle, there exists a function whose graph matches the circle around that point. It is then perfectly valid to find the derivative of that implicit function, which will give us the slope of the tangent line to the circle at that point.

Let's go over the method of implicit differentiation you encountered in single variable calculus, again using a circle.

$$x^2 + y^2 = 25$$

Suppose we want the slope of the tangent line to the circle at $(3,4)$. The implicit function theorem tells us that there exists a function $y = f(x)$ whose graph looks like the circle around $(3,4)$, so it must satisfy the equation

$$x^2 + f(x)^2 = 25$$

We can then differentiate both sides with respect to x, using the chain rule where appropriate.

$$2x + 2f(x)f'(x) = 0$$

$$f'(x) = -\frac{x}{f(x)}$$

Therefore the slope we are after is

$$f'(3) = -\frac{3}{f(3)} = -\frac{3}{4}$$

We can always use this process, but it turns out we can also just find general formulas for the derivatives of these implicit functions using the multivariable chain rule.

Example. Given a curve in $\mathbb{R}^2$ which is the zero set of $G(x,y)$ and a point **a** where the second partial of G is nonzero, find an expression for $f'(x)$, the derivative of the function guaranteed by the implicit function theorem.

We know that since the graph of the function matches the curve, it satisfies $G(x, f(x)) = 0$ around **a**. Let $C(x)$ be the composition of the two functions $G(x,y)$ and $g(x) = \langle x, f(x)\rangle$ so that $C(x) = G(g(x)) = G(x, f(x)) = 0$. We apply the chain rule to this composition function.

$$\left[C'(x)\right] = \begin{bmatrix} \frac{\partial G}{\partial x} & \frac{\partial G}{\partial y} \end{bmatrix} \begin{bmatrix} 1 \\ f'(x) \end{bmatrix}$$

$$C'(x) = \frac{\partial G}{\partial x} + \frac{\partial G}{\partial y}f'(x)$$

But $C(x) = 0$ at the points we are concerned about, so its derivative must be equal to zero, and we have

$$\frac{\partial G}{\partial x} + \frac{\partial G}{\partial y} f'(x) = 0$$

$$\frac{\partial G}{\partial y} f'(x) = -\frac{\partial G}{\partial x}$$

$$f'(x) = -\frac{\partial G/\partial x}{\partial G/\partial y}$$

Using this formula, we could have immediately calculated the slope of the circle $x^2 + y^2 = 25$ at $(3, 4)$. Defining $G(x, y) = x^2 + y^2 - 25$, we would obtain

$$f'(3) = -\frac{\partial G/\partial x}{\partial G/\partial y} = -\frac{2x}{2y} = \frac{3}{4}$$

Example. Consider the curve in $\mathbb{R}^2$ given by the equation $3x^2 + y^3 + xy = 5$. Find the slope of the tangent line at $(1, 1)$ and show that this is possible in the first place. To start with, we represent the curve as the zero set of function

$$G(x, y) = 3x^2 + y^3 + xy - 5$$

We want to show that there exists a function $y = f(x)$ which matches this curve in a neighborhood around $(1, 1)$, so we check that the second partial is nonzero at this point.

$$\frac{\partial G}{\partial y} = 3y^2 + x = 4 \neq 0$$

So the implicit function theorem guarantees the existence of a differentiable function $f(x)$ around $(1, 1)$. We can then compute the derivative using the formula from last example.

$$f'(x) = -\frac{\partial G/\partial x}{\partial G/\partial y}$$

$$= -\frac{6x + y}{3y^2 + x}$$

$$= -\frac{7}{4}$$

Now we look at how the implicit function theorem applies to surfaces in $\mathbb{R}^3$ which can be represented as the zero set of some function $G(x, y, z)$. Often times we want to find z as a function of x and y in an area around some point $\mathbf{a}$. The condition for this, according to the theorem, is that $\frac{\partial G}{\partial z}$ be nonzero at $\mathbf{a}$. If this is true, then we can find a differentiable function $f(x, y)$ whose graph matches the surface locally around $\mathbf{a}$. For example, you can easily see that the top half of the sphere $x^2 + y^2 + z^2 = 1$ can be represented as the graph of

$$f(x, y) = \sqrt{1 - x^2 - y^2}$$

As before, we might want to differentiate this function, and we can find formulas to do that using the chain rule.

Example. Given a surface in $\mathbb{R}^3$ which is the zero set of $G(x, y, z)$ and a point $\mathbf{a}$ where the third partial of G is nonzero, find an expression for $\frac{\partial f}{\partial x}$ and $\frac{\partial f}{\partial y}$, the first partials of the function guaranteed by the implicit function theorem.

As in example 1, let $C(x, y)$ be the composition of the two functions $G(x, y, z)$ and $g(x, y) = \langle x, y, f(x, y) \rangle$ and apply the chain rule.

$$\begin{bmatrix} \dfrac{\partial C}{\partial x} & \dfrac{\partial C}{\partial y} \end{bmatrix} = \begin{bmatrix} \dfrac{\partial G}{\partial x} & \dfrac{\partial G}{\partial y} & \dfrac{\partial G}{\partial z} \end{bmatrix} \begin{bmatrix} 1 & 0 \\ 0 & 1 \\ \dfrac{\partial f}{\partial x} & \dfrac{\partial f}{\partial y} \end{bmatrix}$$

$$\frac{\partial C}{\partial x} = \frac{\partial G}{\partial x} + \frac{\partial G}{\partial z} \frac{\partial f}{\partial x}$$

$$\frac{\partial C}{\partial y} = \frac{\partial G}{\partial y} + \frac{\partial G}{\partial z} \frac{\partial f}{\partial y}$$

Since the function $f(x, y)$ satisfies $G(x, y, f(x, y)) = 0$ around $\mathbf{a}$, we have $C(x, y) = 0$ so its partial derivatives must be zero. Then

$$\frac{\partial f}{\partial x} = -\frac{\partial G / \partial x}{\partial G / \partial z}$$

$$\frac{\partial f}{\partial y} = -\frac{\partial G / \partial y}{\partial G / \partial z}$$

As a quick example to illustrate this, consider the unit sphere $x^2 + y^2 + z^2 = 1$ which can be represented as the zero set of $G(x, y, z) = x^2 + y^2 + z^2 - 1$. At any point where $z \neq 0$, the implicit function theorem guarantees the existence of a function $f(x, y)$ whose graph matches the sphere around the point. By the formulas we just found,

$$\frac{\partial f}{\partial x} = -\frac{x}{z}$$

$$\frac{\partial f}{\partial y} = -\frac{y}{z}$$

Gradient Vectors and Level Sets

With the implicit function theorem, we can now prove another useful property of gradient vectors that we didn't cover in section 3.3.

Suppose we are standing at a point (x_0, y_0) on the zero set of some function $G(x, y) = 0$. Now imagine we move along some vector $\mathbf{v}$, where $\mathbf{v}$ points along the level set we are currently on. In other words, $\mathbf{v}$ is *tangent* to this curve at (x_0, y_0).

What will the directional derivative of G with respect to $\mathbf{v}$ be at this point? We know intuitively that it should be zero. By definition, the value of a function does not change as we move along one of its level sets, so the rate of change given by the directional derivative must be 0. Assuming that g is a differentiable function, this means

$$\nabla G(x_0, y_0) \cdot \mathbf{v} = 0$$

So the gradient vector of G at this point is *orthogonal* to $\mathbf{v}$, a vector which is tangent to the level set $G(x, y) = 0$, and is therefore orthogonal to the level set there.

Since we can go through the same argument for any point along any level set of G, we conclude that the gradient vector $\nabla G(x, y)$ is always orthogonal to any level set of G.

To make this a little more rigorous, we can utilize the implicit function theorem from last section. We basically need this to prove that a tangent vector such as $\mathbf{v}$ actually exists.

As long as one of the partial derivatives of G is nonzero at (x_0, y_0), the theorem guarantees that we can solve for y as a function of x or vice versa, and that the graph of this implicit function matches the level set of G around that point.

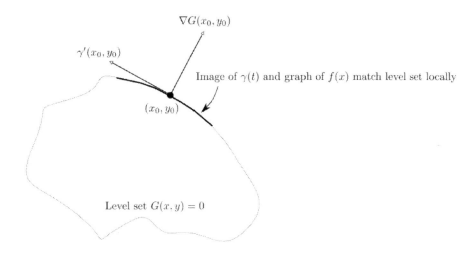

Remember that if we have the graph of a function, we can always construct a vector function whose image matches that graph. For example, if our implicit function is $f(x)$, then we can make a function $\gamma(t) = \langle t, f(t) \rangle$. This parametric curve will then locally match the graph of the implicit function and therefore the original level set $G(x, y)$ as well, i.e.

$$G(\gamma(x_0)) = 0$$

for all points in a neighborhood around (x_0, y_0). By the chain rule,

$$\nabla G(x_0, y_0) \cdot \gamma'(x_0) = 0$$

This tells us that the gradient is orthogonal to the derivative vector of this curve at (x_0, y_0). We know that the derivative of a vector function points tangent to its image, so $\gamma'(x_0)$ must be tangent to the level set and therefore $\nabla G(x_0, y_0)$ is orthogonal to the level set.

While we have only been talking about a level set in two dimensional space, these facts generalize to any level set. The gradient vector of a function is always orthogonal to its level sets.

Exercises

1. Consider the curve given by $2x^3 + y^3 = 5xy$

    (a) Show that a differentiable function $f(x)$ exists whose graph matches the curve locally around the point $(1, 2)$

    (b) Find the slope of the tangent line to the curve at $(1, 2)$

2. Consider the surface given by $2z^3 + 3y^5 + e^{x+z} = 25$

    (a) Show that at every point on the surface, z can be viewed locally as a function of x and y

    (b) Find formulas for the partial derivatives of z with respect to y and x

3. Find an equation for the tangent plane to the ellipsoid $x^2 + 4y^2 + z^2 = 6$ at the point $(1, 1, 1)$. (Hint: utilize the fact that the gradient vector of a function is orthogonal to its level sets)

Answers

1. (a) Let $G(x, y) = 2x^3 + y^3 - 5xy$; then

$$\frac{\partial G}{\partial y} = 3y^2 - 5x = 7 \neq 0$$

    (b) $f'(1) = \frac{4}{7}$

2. (a) Let $G(x, y, z) = 2z^3 + 3y^5 + e^{x+z} - 25$; then

$$\frac{\partial G}{\partial z} = 6z^2 + e^{x+z} \neq 0$$

    because the first term is never negative and the second is always positive

    (b)

$$\frac{\partial z}{\partial x} = -\frac{e^{x+z}}{6z^2 + e^{x+z}}$$

$$\frac{\partial z}{\partial y} = -\frac{15y^4}{6z^2 + e^{x+z}}$$

3. The ellipsoid is the zero set of $G(x, y, z) = x^2 + 4y^2 + z^2 - 6$. To describe the tangent plane, we need a point and a normal vector. The gradient $\nabla G(1, 1, 1)$ must be orthogonal to the ellipsoid at $(1, 1, 1)$ since the ellipsoid is a level set of G, therefore we can choose the gradient as the normal vector.

$$(x - 1) + 4(y - 1) + (z - 1) = 0$$

3.7 Critical Points

In the next few sections, we will be concerned with finding maxima and minima of multivariable functions. We have already seen that a point at which the gradient vector equals the zero vector is a good candidate for a local max or min because it means there is no direction of fastest increase or decrease there. However, the problem is these points, called **critical points**, can be neither local maxima nor minima, in which case they are called **saddle points**.

For a single variable function, we could determine if a critical point a was a max or min by using either the first derivative test or the second derivative test. The first derivative test involved finding out if the value of the derivative switched signs around the point a, while the second derivative test involved finding out if the second derivative at a was positive or negative. The first derivative test does not work at all for multivariable functions, but it turns out we can develop a more complicated version of the second derivative test which does work.

This generalization of the second derivative test will be our goal for this section. An actual proof of the test involves a whole lot of linear algebra, but we can develop a somewhat convincing argument using only familiar quadratic polynomials.

Before we do anything, however, we should go over what we mean by a local max or min. A function has a *local maximum* at $\mathbf{a}$ if $f(\mathbf{a})$ is greater than *or equal to* $f(\mathbf{x})$ for all $\mathbf{x}$ in some neighborhood around $\mathbf{a}$. Likewise, a function has a *local minimum* at $\mathbf{a}$ if $f(\mathbf{a})$ is less than *or equal to* $f(\mathbf{x})$ for all $\mathbf{x}$ in some local area surrounding $\mathbf{a}$. So as you'd expect, a local max or min is an extreme value only in its immediate surroundings.

Quadratics

Our first step is to examine quadratic polynomials of the form $ax^2 + 2bx + c$. For reasons which you will soon see, we want to know whether this polynomial is always positive, always negative, or neither. If the quadratic is positive for all values of x, we will call it *positive definite*. If it is always negative, we say it is *negative definite*, and if it takes both positive and negative values it is called *nondefinite*.

The easiest way to find out if this quadratic is positive, negative, or nondefinite is to think of it as a parabola in the xy plane. We can find the roots of the parabola by the good old quadratic equation:

$$x = \frac{-2b \pm \sqrt{4b^2 - 4ac}}{2a}$$

$$x = \frac{-b \pm \sqrt{b^2 - ac}}{a}$$

The sign of the stuff in the square root, called the discriminant, tells us if the roots of the parabola are real or imaginary.

If the discriminant is negative ($b^2 - ac < 0$), then the roots are imaginary and the parabola never touches the x axis, so it must be always positive or always negative. We can determine which by looking at the sign of a, the coefficient of the squared term. If this term is positive, the parabola opens upward and therefore must lie completely above the x axis. Conversely, if a is negative, the parabola opens downward and stays under the x axis.

If the discriminant is positive ($b^2 - ac > 0$), then the parabola has two real roots, so it crosses the x axis twice and must take on both positive and negative values in the process.

Lastly, if the discriminant is zero ($b^2 - ac = 0$), then the parabola has one real root of multiplicity 2, meaning it kind of touches the x axis then bounces off. In this case we cannot say it is always positive or always negative since it reaches 0 at its root. However, we also cannot say it has both positive and negative values.

We rearrange these inequalities and sum up our findings:

A quadratic polynomial $ax^2 + 2bx + c$ is:

- positive definite if $ac - b^2 > 0$ and $a > 0$

- negative definite if $ac - b^2 > 0$ and $a < 0$

- nondefinite if $ac - b^2 < 0$

- none of the above if $ac - b^2 = 0$

Taylor Polynomials

Remember that a Taylor polynomial of some function $f(x)$ centered at a provides an approximation to the function around that point. In particular, the second degree Taylor polynomial centered at a looks like this:

$$f(a + h) \approx f(a) + f'(a)h + \frac{1}{2}f''(a)h^2$$

where h is how far away we are from a, in other words our displacement from the initial point. If a happens to be a critical point of the function, then the first derivative there is zero and the Taylor polynomial becomes

$$f(a+h) \approx f(a) + \frac{1}{2}f''(a)h^2$$

To determine if the function achieves a local max or min at a, we must look at the value of the function at points surrounding a. By the Taylor polynomial, we see that these values depend on the sign of the second derivative.

If $f''(a) > 0$, then for all h, $f(a+h)$ must be greater than or equal to $f(a)$. Therefore, the function has a local minimum at a. Similarly, if $f''(a) < 0$, then $f(a+h)$ is always less than or equal to $f(a)$ so there is a local maximum. If $f''(a) = 0$, then we cannot conclude anything: higher derivatives would need to be considered. This, of course, is the second derivative test you learned in single variable calculus.

To apply this test to multivariable functions, we need to introduce multivariable Taylor polynomials. These can get very complicated, but luckily we only need the second degree approximation for our purposes.

The basic idea of the multivariable Taylor polynomial is the same: we want an approximation to the function centered at $\mathbf{a}$ so that when we move with displacement $\mathbf{h}$, we can predict what happens to the function. The second degree Taylor polynomial for a real valued function $f(x,y)$ looks like this

$$f(\mathbf{a}+\mathbf{h}) \approx f(\mathbf{a}) + D_{\mathbf{h}}f(\mathbf{a}) + \frac{1}{2}D_{\mathbf{h}}^2 f(\mathbf{a})$$

The $D_{\mathbf{h}}^2$ indicates taking the directional derivative twice, i.e. we take the directional derivative of f with respect to $\mathbf{h}$, then take the directional derivative of that with respect to $\mathbf{h}$.

If $\mathbf{a}$ is a critical point, then $\nabla f(\mathbf{a}) = \mathbf{0}$ and the directional derivative must be zero as well, so the polynomial becomes

$$f(\mathbf{a}+\mathbf{h}) \approx f(\mathbf{a}) + \frac{1}{2}D_{\mathbf{h}}^2 f(\mathbf{a})$$

As before, to determine whether the function has a local max, min, or saddle point at $\mathbf{a}$, we need to look at the values of the function around $\mathbf{a}$. As you can see, whether the function assumes values greater or less than $f(\mathbf{a})$ when we move along $\mathbf{h}$ depends on the sign of $D_{\mathbf{h}}^2 f(\mathbf{a})$. We will now see that this actually turns out to be a quadratic expression. Taking the directional derivative one time gets us

$$D_{\mathbf{h}}f(x,y) = \nabla f(x,y) \cdot \mathbf{h} = \frac{\partial f}{\partial x}h_1 + \frac{\partial f}{\partial y}h_2$$

To take the directional derivative again, we find the gradient of this and then dot it with **h** again.

$$\nabla D_{\mathbf{h}} f(x,y) = \left\langle \frac{\partial^2 f}{\partial x^2} h_1 + \frac{\partial^2 f}{\partial x \partial y} h_2, \frac{\partial^2 f}{\partial y \partial x} h_1 + \frac{\partial^2 f}{\partial y^2} h_2 \right\rangle$$

The expression we are after is then

$$
\begin{aligned}
D_{\mathbf{h}}^2 f(x,y) &= \nabla D_{\mathbf{h}} f(x,y) \cdot \mathbf{h} \\
&= \left(\frac{\partial^2 f}{\partial x^2} h_1 + \frac{\partial^2 f}{\partial x \partial y} h_2 \right) h_1 + \left(\frac{\partial^2 f}{\partial y \partial x} h_1 + \frac{\partial^2 f}{\partial y^2} h_2 \right) h_2 \\
&= \frac{\partial^2 f}{\partial x^2} h_1^2 + 2 \frac{\partial^2 f}{\partial x \partial y} h_1 h_2 + \frac{\partial^2 f}{\partial y^2} h_2^2 \\
&= h_2^2 \left[\frac{\partial^2 f}{\partial x^2} \left(\frac{h_1}{h_2} \right)^2 + 2 \frac{\partial^2 f}{\partial x \partial y} \left(\frac{h_1}{h_2} \right) + \frac{\partial^2 f}{\partial y^2} \right]
\end{aligned}
$$

Since h_2^2 is always positive, the sign of this thing only depends on the sign of

$$\frac{\partial^2 f}{\partial x^2} \left(\frac{h_1}{h_2} \right)^2 + 2 \frac{\partial^2 f}{\partial x \partial y} \left(\frac{h_1}{h_2} \right) + \frac{\partial^2 f}{\partial y^2}$$

where all of these partial derivatives are evaluated at **a**.

But this is a quadratic polynomial of the form $ax^2 + 2bx + c$, if we think of (h_1/h_2) as x. The partial derivatives (which are evaluated at **a**) in front of each term correspond to the constants a, b, and c.

We just saw how to tell if these kinds of quadratics are positive definite, negative definite, nondefinite, or neither. Looking back at the Taylor polynomial,

$$f(\mathbf{a} + \mathbf{h}) \approx f(\mathbf{a}) + \frac{1}{2} D_{\mathbf{h}}^2 f(\mathbf{a})$$

we see that if this quadratic is positive definite, then $f(\mathbf{a} + \mathbf{h})$ is always greater than or equal to $f(\mathbf{a})$, so the function has a local minimum.

If it is negative definite, there is a local maximum since the function takes on values less than or equal to $f(\mathbf{a})$ close by. If it is nondefinite, the function takes on values both greater than and less than $f(\mathbf{a})$ around it, so it cannot be a max or min and must be a saddle point. If it is none of the above, we don't know what happens; you would have to look at a higher order Taylor polynomial. This is the **_second derivative test_** for two variable functions.

Theorem. Suppose $f(x,y)$ is a differentiable, real valued function and **a** is a critical point of f. Let

$$\Delta = ac - b^2 = \frac{\partial^2 f}{\partial x^2} \frac{\partial^2 f}{\partial y^2} - \left(\frac{\partial^2 f}{\partial x \partial y} \right)^2$$

In the following statements, Δ and any partial derivatives of f are evaluated at **a**.

- if $\Delta > 0$ and $\dfrac{\partial^2 f}{\partial x^2} > 0$, then the function has a local min at $\mathbf{a}$

- if $\Delta > 0$ and $\dfrac{\partial^2 f}{\partial x^2} < 0$, then the function has a local max at $\mathbf{a}$

- if $\Delta < 0$, then $\mathbf{a}$ is a saddle point

- if $\Delta = 0$, this test is useless.

Example. Find and classify any critical points of $f(x, y) = 4x^2 + y^2 + 3x$. We find points where the gradient vector is the zero vector

$$\nabla f(x, y) = \langle 8x + 3, 2y \rangle$$

$$8x + 3 = 0 \qquad \text{and} \qquad 2y = 0$$

The only critical point seems to be $(-\frac{3}{8}, 0)$. We apply the second derivative test at this point.

$$\Delta = \frac{\partial^2 f}{\partial x^2} \frac{\partial^2 f}{\partial y^2} - \left(\frac{\partial^2 f}{\partial x \partial y} \right)^2$$

$$= (8)(2) - (0)^2 = 16$$

We see that $\Delta > 0$ and $\dfrac{\partial^2 f}{\partial x^2} > 0$, so $f(x, y)$ has a local minimum at $(-\frac{3}{8}, 0)$

Exercises

1. Find and classify the critical points of the function

   (a) $f(x, y) = x^3 - 3x + y^2$

   (b) $f(x, y) = xy + 2x - y$

2. Show that the second degree Taylor polynomial of $f(x, y) = e^{x+y}$ centered at $(0, 0)$ is

   $$T(\mathbf{h}) = 1 + (h_1 + h_2) + \frac{1}{2}(h_1^2 + 2h_1 h_2 + h_2^2)$$

   (a) Use this to approximate $e^{.3}$

Answers

1. (a) $(1, 0)$ local min; $(-1, 0)$ saddle point

   (b) $(1, -2)$ saddle point

2. Follow the same procedure shown in the section, then plug in 1 for all the partial derivatives

(a) Using a displacement vector such as $\mathbf{h} = \langle .1, .2 \rangle$,

$$e^{.3} \approx T(\mathbf{h}) = 1.345$$

3.8   Lagrange Multipliers

Critical points are nice and all, but sometimes we want to maximize or minimize a function over a certain constraint, not on the whole domain. For example, say we want to find the point on the unit circle $x^2 + y^2 = 1$ which is closest to the point $(.2, .5)$.

Wording this as a max/min problem, we want to minimize the distance function

$$f(x, y) = (x - .2)^2 + (y - .5)^2$$

(we square it so we don't have to deal with square roots) on the circle $x^2 + y^2 = 1$. The circle is a *constraint* on the function; we are only interested in max/min points that are also on this circle. If we just found the critical points of f, we would obviously find that the point $(.2, .5)$ itself minimizes the distance from the point $(.2, .5)$. So it seems that we need a new method for finding max/min points when optimizing a function subject to a constraint.

Two Dimensions

Here is the problem: we want to maximize or minimize a function $f(x, y)$ subject to some constraint, which we will express as the zero set of some function $G(x, y) = 0$. So we are only interested in a max/min point $\mathbf{a}$ if it is also on this level set, i.e. $G(\mathbf{a}) = 0$. If $\mathbf{a}$ is a point on the constraint where f achieves a local max or min, we say $\mathbf{a}$ is a *constrained critical point*. Note that when we say local max or min in this context, we mean an extreme value compared to all points around it *which are also on the constraint.*

It turns out that at all constrained critical points, the gradient vectors of f and G must be parallel. In others words, $\nabla f(\mathbf{a})$ and $\nabla G(\mathbf{a})$ are scalar multiples of each other when $\mathbf{a}$ is a constrained critical point.

Theorem. Suppose the function $f(x, y)$ obtains a local maximum or minimum value on the level curve $G(x, y) = 0$ at the point $\mathbf{a}$. If $\nabla G(\mathbf{a}) \neq \mathbf{0}$, then the following must be true:

$$\nabla f(\mathbf{a}) = \lambda \nabla G(\mathbf{a})$$

for some real number λ, which is called a **_Lagrange multiplier_**.

Proof. All we need to prove is that both ∇f and ∇G are orthogonal to the level set $G(x, y) = 0$ at $\mathbf{a}$. Once we prove this, it follows that the gradients are parallel because in two dimensional space if two vectors are both orthogonal to a third vector, they must be parallel to each other.

We know that $\nabla G(\mathbf{a})$ is perpendicular to the constraint curve if $\nabla G(\mathbf{a}) \neq \mathbf{0}$ because gradients of a function are orthogonal to any level set of the function. If $\nabla G(\mathbf{a}) = \mathbf{0}$, then we cannot use this fact because its proof involved the implicit function theorem (section 3.6), which does not hold when the gradient is the zero vector.

To show that the gradient of f will be orthogonal to the level set, imagine a curve $\gamma(t)$ which matches the constraint curve around $\mathbf{a}$ such that $\gamma(t_0) = \mathbf{a}$. The existence of this curve is guaranteed by the implicit function theorem, assuming that $\nabla G(\mathbf{a}) \neq \mathbf{0}$. Then the derivative vector $\gamma'(t_0)$ points tangent to the level set at $\mathbf{a}$.

Since $f(x, y)$ reaches a max/min at $\mathbf{a}$, it follows that a composition function $h(t) = f(\gamma(t))$ reaches a max/min at the point $t = t_0$ because $h(t_0) = f(\gamma(t_0)) = f(\mathbf{a})$. Therefore, the derivative of h at this point is zero, and by the chain rule,

$$h'(t_0) = \nabla f(\mathbf{a}) \cdot \gamma'(t_0) = 0$$

So $\nabla f(\mathbf{a})$ is orthogonal to a vector which is tangent to the level set, and therefore it is orthogonal to the constraint at $\mathbf{a}$. $\qquad \square$

The general idea is that the Lagrange multiplier condition will simplify into a system of three equations:

$$\frac{\partial f}{\partial x} = \lambda \frac{\partial G}{\partial x}$$
$$\frac{\partial f}{\partial y} = \lambda \frac{\partial G}{\partial y}$$
$$G(x, y) = 0$$

which can be solved for the three unknowns x, y, and λ. Any point (x, y) which satisfies all three equations for some value of λ is a constrained critical point. Finding these points can be very tricky sometimes because most of the time the equations will not be a linear system of equations which you are used to solving.

One general rule to keep in mind when solving these equations is to never divide by x or y unless you are certain that they cannot be zero, or if you have already solved the cases when they are zero. Once you find all the constrained critical points, plug them all into the original function $f(x, y)$ and find out which points yield the maximum and minimum values.

Example. Find the points on the circle $x^2 + y^2 = 4$ which are closest to and farthest from the point $(1, 1)$. So we want to optimize the function $f(x, y) = (x - 1)^2 + (y - 1)^2$ on the constraint $G(x, y) = x^2 + y^2 - 4 = 0$. First, we simply write out the Lagrange multiplier condition

$$\nabla f(x, y) = \lambda \nabla G(x, y)$$

$$\langle 2(x-1), 2(y-1) \rangle = \lambda \langle 2x, 2y \rangle$$

So our system of three equations is

$$2(x-1) = \lambda 2x$$

$$2(y-1) = \lambda 2y$$

$$x^2 + y^2 = 4$$

We focus on the first two equations. Dividing by 2 is legal, so we do that. Notice that the right sides of both equations look very similar: if we multiply the first equation by y and the second equation by x, they will be the same.

$$y(x-1) = \lambda xy = x(y-1)$$

$$xy - y = xy - x$$

$$x = y$$

So we have taken the first two equations and reduced them down to the requirement that $x = y$. Now we can use the third equation by replacing y by x or vice versa, then solving.

$$x^2 + y^2 = x^2 + x^2 = 2x^2 = 4$$

$$x = \pm\sqrt{2}$$

Therefore, our possible max/min points are $(\sqrt{2}, \sqrt{2})$ and $(-\sqrt{2}, -\sqrt{2})$. If you plug these points into $f(x, y)$ you will find that

$$f(\sqrt{2}, \sqrt{2}) \approx .343$$

$$f(-\sqrt{2}, -\sqrt{2}) \approx 11.66$$

So the points minimize and maximize, respectively, the distance from the point $(1, 1)$ when we are constrained to the circle $x^2 + y^2 = 4$. If you draw a picture of this situation, it makes a lot of sense.

Three Dimensions

The method of Lagrange multipliers extends to three dimensions by an argument similar to the one we used in the two dimensional case.

Suppose we want to optimize the function $f(x, y, z)$ on the surface represented by the level set $G(x, y, z) = 0$. If $\mathbf{a}$ is a constrained critical point, then we again have

$$\nabla f(\mathbf{a}) = \lambda \nabla G(\mathbf{a})$$

for some number λ, provided that $\nabla G(\mathbf{a}) \neq \mathbf{0}$.

We know that the gradient of g will always be orthogonal to the level set, i.e. it will be orthogonal to the tangent plane to the surface at $\mathbf{a}$. We can deduce that the gradient of f will also be orthogonal to this tangent plane at $\mathbf{a}$ by using curves that run through the surface. If $\gamma(t)$ is a curve on the level surface such that $\gamma(t_0) = \mathbf{a}$, you can show that $\nabla f(\mathbf{a})$ is orthogonal to the derivative vector $\gamma'(t_0)$ by the same procedure used in the two variable case. Since this argument holds for *every* possible curve, then $\nabla f(\mathbf{a})$ is orthogonal to all these tangent vectors to the surface at $\mathbf{a}$, so it is orthogonal to the tangent plane there. Then the gradients of f and g are both orthogonal to the tangent plane, and therefore must be parallel.

Example. Find the rectangular box with a volume of 1000 and dimensions x, y, and z that minimize the total surface area. We want to minimize the function $f(x, y, z) = 2xy + 2xz + 2yz$ on the constraint $G(x, y, z) = xyz - 1000 = 0$. The Lagrange multiplier condition gives us a system of four equations:

$$2y + 2z = \lambda yz$$

$$2x + 2z = \lambda xz$$

$$2x + 2y = \lambda xy$$

$$xyz = 1000$$

We use the same trick from last example on the first three equations. If we multiply the first, second, and third equations by x, y, and z, respectively, then the right sides will all be equal.

$$2xy + 2xz = 2xy + 2yz = 2xz + 2yz = \lambda xyz$$

From this you can easily deduce

$$2xy + 2xz = 2xy + 2yz \qquad \text{and} \qquad 2xy + 2yz = 2xz + 2yz$$

$$xz = yz \qquad \text{and} \qquad xy = xz$$

$$x = y \qquad \text{and} \qquad y = z$$

because you can safely divide by a variable in this problem (if one of the dimensions was 0, then there would not be a volume of 1000). So we know that the dimensions are all equal (the box is a cube). Plugging this into the fourth equation yields

$$xyz = x^3 = 1000$$

$$x = y = z = 10$$

Therefore, a cube with sides of 10 gives the rectangular box with 1000 volume and minimum surface area.

Exercises

1. Find the point on the plane $x + 3y + z = 1$ closest to the origin

2. Find the point on the parabola $y^2 = 3x$ which is closest to the point $(1,0)$

3. Find the three dimensional vector with magnitude 6 whose components have the largest possible sum

Answers

1. $(\frac{1}{11}, \frac{3}{11}, \frac{1}{11})$; minimize $f(x,y,z) = x^2 + y^2 + z^2$ on the constraint $G(x,y,z) = x + 3y + z - 1$

2. $(0,0)$; minimize $f(x,y) = (x-1)^2 + y^2$ on the constraint $G(x,y) = y^2 - 3x$; the system is

$$2(x-1) = -3\lambda$$

$$y = \lambda y$$

$$y^2 = 3x$$

   If y is not zero, then λ must be 1, but this yields $x = -1/2$ which is impossible. Therefore, $y = 0$ which implies $x = 0$

3. $\langle \sqrt{12}, \sqrt{12}, \sqrt{12} \rangle$; maximize $f(x,y,z) = x + y + z$ on the constraint $G(x,y,z) = x^2 + y^2 + z^2 - 36$

3.9   Optimization

In the previous two sections, we developed methods to find points where a function might reach a maximum or minimum. If there is no particular restraint on f, then we find local extrema by looking at the critical points of f, which we can classify with the second derivative test. If we are only looking at the behavior of the function along some constraint curve or surface, then we use Lagrange multipliers. In this section, we will put both of these methods together to optimize functions on a special type of domain.

Closed and Bounded Sets

We are interested specifically in optimizing functions over a domain which is *closed* and *bounded*. A detailed discussion of these terms would take too long, so we will look at their intuitive definitions.

Say we want to optimize a function $f(x, y)$ over a domain which looks like the unit circle. If I said to optimize the function over the domain $x^2 + y^2 \leq 1$ and over the domain $x^2 + y^2 < 1$, what is the difference? The second domain is the unit circle without its *boundary*.

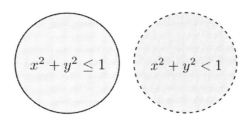

When a set includes its boundary, we say that it is **closed**. So the domain defined by $x^2 + y^2 \leq 1$ would be a closed set, while the other circle would not be. As another example, consider an interval on the real number line, which could be the domain of a single variable function. A closed interval is a closed set because it contains its boundary, which is just two points, e.g. the boundary of $[2, 4]$ is the points 2 and 4. An open interval $(2, 4)$ is not a closed set. For a three dimensional domain, the boundary would be a two dimensional surface, e.g. the unit ball with the unit sphere would be a closed set because the sphere is the boundary of the ball.

A set is said to be **bounded** if it fits inside some other finite set. Basically, this just means the set doesn't go off to infinity in some direction. For example, the interval on $\mathbb{R}$ $(2,4)$ would be bounded because we can fit it inside another interval, such as $(0,5)$. The unit circle in $\mathbb{R}^2$ is a bounded set because we can fit it inside another two dimensional set, such as a circle of radius 2.

Global Extrema

Suppose we have a continuous function f defined on a *closed* and *bounded* set D. Then f must attain maximum and minimum values over the set D. This is a very important statement which we can't prove without going off topic for a bit.

We will call the maximum and minimum points on D guaranteed by this statement **global extrema**. A function has a *global maximum* at a point $\mathbf{a}$ if $f(\mathbf{a})$ is greater than or equal to the value of the function *at all points in* D. Similarly, a function has a *global minimum* at $\mathbf{a}$ if $f(\mathbf{a}) \leq f(\mathbf{x})$ for all $\mathbf{x}$ in the closed and bounded domain D.

Optimizing

Now that we have some terminology established, we are ready to start optimizing functions over closed and bounded sets. The procedure is somewhat similar to the one used in single variable calculus. Remember that if we had a function $f(x)$ over a closed interval $[a, b]$, then the function had to have a global max and global min on this interval (since it is a closed and bounded set).

To find possible points at which these occurred, we needed to do two things. First, we checked for any critical points of the function that were inside this interval. Next, we looked at the boundary of the interval, which were just the two end points a and b. We then tested all these points to see where the global max and min occurred.

For a two or three variable function over a closed and bounded set, the procedure is as follows. First, we check if any critical points of f are inside the domain we are optimizing over. Next, we use the method of Lagrange multipliers to check for possible max and min points on the boundary of the domain (which will consist of one or more constraints). Sometimes, however, one part of the boundary is simple enough that we do not have to use Lagrange multipliers. Lastly, we gather all these candidates and plug them into the function to find the global max and min.

Example. Find the global maximum and minimum of the function $f(x, y) = 3xy - 6x - 3y + 7$ on the closed triangle in the xy plane with vertices at $(0, 0)$, $(0, 5)$, and $(3, 0)$. First, we find any critical points which lie in the *interior* (not on the boundary) of the triangle domain.

$$\nabla f(x, y) = \langle 3y - 6, 3x - 3 \rangle$$

We see that the only critical point occurs at $(1, 2)$ which does lie inside our domain, so we add that to our list of points to check and move on to the boundary.

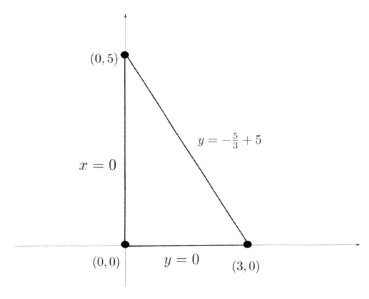

We can split the boundary into 6 different parts which we must check separately: the line segment $y = 0$ from $0 < x < 3$, the line segment $x = 0$ from $0 < y < 5$, the line $y = -\frac{5}{3} + 5$ between the points $(0, 5)$ and $(3, 0)$, and finally the three vertices of the triangle.

First let's check along $y = 0$. We could use Lagrange multipliers, but that would be overly complicated. Instead, we will look at the behavior of our function when $y = 0$. In other words, convert f to a single variable function by plugging in $y = 0$, then find the critical points like you did in single variable calculus.

$$f(x, 0) = -6x + 7$$

$$f'(x, 0) = -6$$

The derivative is never 0, so we have no constrained critical points along the line $y = 0$ from $0 < x < 3$. We follow a similar procedure for the line segment $x = 0$: convert the function to a single variable function by plugging in $x = 0$ and find the critical points.

$$f(0, y) = -3y + 7$$

$$f'(0, y) = -3$$

Again, there are no constrained critical points on the line segment $x = 0$ from $0 < y < 5$.

For the third line segment, $y = -\frac{5}{3} + 5$, we will illustrate how to apply Lagrange multipliers. You could, of course, still do the same procedure as for the other two line segments: replace y with $-\frac{5}{3}x + 5$ in the original function. The constraint function is the zero set of $G(x, y) = y + \frac{5}{3}x - 5$.

$$\nabla f(x, y) = \lambda \nabla G(x, y)$$

$$\langle 3y - 6, 3x - 3 \rangle = \lambda \langle \frac{5}{3}, 1 \rangle$$

Our system of three equations is therefore

$$3y - 6 = \lambda \frac{5}{3}$$

$$3x - 3 = \lambda$$

$$y + \frac{5}{3}x = 5$$

One way to solve this is to multiply the first equation by $\frac{3}{5}$, then set the first and second equations equal to each other because their right sides will both be λ.

$$\frac{9}{5}y - \frac{18}{5} = 3x - 3$$

$$x = \frac{3}{5}y - \frac{1}{5}$$

Now we plug this into the third equation, solve for y, then use that to get x.

$$y + \frac{5}{3}(\frac{3}{5}y - \frac{1}{5}) = 5$$

$$y = \frac{8}{3} \qquad \text{and} \qquad x = \frac{7}{5}$$

So we have one critical point on the interior, one constrained point along the boundary line segment $y = -\frac{5}{3}x + 5$, and three vertices to check. To convince yourself that we need to check the vertices separately, try to think about what it would mean if a vector is orthogonal to a single point. This notion doesn't really make sense, and since Lagrange multipliers are based on gradient vectors being orthogonal to your constraint, we must check them individually.

$$f(1,2) = 1 \qquad f(\frac{7}{5}, \frac{8}{3}) = \frac{9}{5}$$

$$f(0,0) = 7 \qquad f(3,0) = -11 \qquad f(0,5) = -8$$

So both the global max and min occur at vertices, and we wasted a bunch of time finding other possible points.

Exercises

1. Describe the maxima and minima of the function $z = 3$

2. Optimize the function $f(x,y) = xy - 2x$ over the triangle in the xy plane with vertices $(0,0)$, $(0,4)$, $(4,0)$

3. Find the global maximum of $f(x,y,z) = x^2 - y^2 + z^2$ on the unit ball

Answers

1. All points are both local and global max and min

2. max: 1; min: -8

3. 1; just by looking at the function, you know $y = 0$ at the maximum; when $y = 0$, $x^2 + z^2 \leq 1$

Chapter 4

Integral Calculus

4.1   Multiple Integrals

You are probably very familiar with the single variable definite integral by now. It is most often learned as the area under the graph of some function $f(x)$ between $x = a$ and $x = b$. While this is certainly correct, we know that the integral has many uses besides just finding the area under a curve. So, in general, what is an integral?

The easiest way to interpret the definite integral is in terms of Riemann sums. Suppose we have the integral $\int_a^b f(x)\,dx$, so we are integrating the function $f(x)$ (the integrand) over some domain. For single variable integrals, the domain we are integrating over is a one dimensional interval $[a, b]$ of the x axis. We start by dividing this interval into n smaller intervals which have the following end points:

$$a = x_0 < x_1 < x_2 < \ldots < x_{n-1} < x_n = b$$

$$[x_0, x_1], [x_1, x_2], \ldots\ldots, [x_{n-1}, x_n]$$

This is called taking a *partition* of the interval $[a, b]$. When talking about a random subinterval in this partition, we will often say the ith interval. So $i = 1$ refers to the first subinterval $[x_0, x_1]$, $i = n$ refers to the last subinterval $[x_{n-1}, x_n]$, etc.

Let Δx_i represent the length of the ith subinterval. We pick a random value inside this ith subinterval, which we will denote by x_i^*, then find the value of our integrand function at this point $f(x_i^*)$. Lastly, we multiply together $f(x_i^*)$ and Δx_i. We repeat this process for all n subintervals, then add the results together.

$$\sum_{i=1}^n f(x_i^*)\Delta x_i = f(x_1^*)\Delta x_1 + f(x_2^*)\Delta x_2 + \ldots + f(x_{n-1}^*)\Delta x_{n-1} + f(x_n^*)\Delta x_n$$

This kind of sum, where we have the size of some interval being multiplied by the value of a function at a random point in that interval, is called a *Riemann sum*. Thinking back to area under the curve $f(x)$, each $f(x_i^*)\Delta x_i$ is just the area of a thin rectangle whose base is the ith subinterval and whose height is the value of the graph there. So when we add the areas of these thin rectangles up, we get an approximation to the area under the curve.

If we increase the number of subintervals, each individual rectangle will get thinner and our approximation to the area under the curve will become more accurate. This motivates us to define the definite integral as the limit as $n \to \infty$ of the Riemann sum.

$$\int_a^b f(x)\,dx = \lim_{n \to \infty} \sum_{i=1}^n f(x_i^*)\Delta x_i$$

So essentially an integral is the sum of many terms, where each term is the product of the size of a subinterval Δx_i and the value of the integrand function at a point in that subinterval $f(x_i^*)$. This way of defining an integral has nothing to do with area under the curve, which is just one of many things we can compute with integrals.

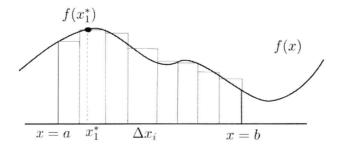

Double Integrals

Now we want to extend the idea of an integral to higher dimensional domains and integrands. The obvious next step would be a two variable integrand $f(x,y)$ over a two dimensional domain in the xy plane, in particular a rectangle. The setup is analogous to the single variable integral.

Our domain of integration will be a rectangle D where $a \le x \le b$ and $c \le y \le d$. We form a partition of both the x and y axes; in other words, we split the intervals $[a,b]$ and $[c,d]$ into many subintervals. Notice that this splits up the rectangle D into a bunch of mini-rectangles determined by the partitions of x and y. Let A_{ij} be the rectangle whose sides are determined by the ith x subinterval and the jth y subinterval (we will use i to refer to random subintervals of x and j to refer to random subintervals of y).

The size of each A_{ij}, denoted by ΔA_{ij} will be the area of the mini-rectangle, which is obtained by multiplying the lengths of its sides $\Delta x_i \Delta y_j$. Next, we pick a random point in this mini-rectangle (x_i^*, y_j^*) and find the value of the integrand function at this point

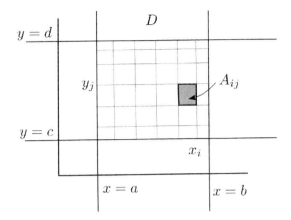

$f(x_i^*, y_j^*)$. Finally, we multiply this by the size of the mini-rectangle, repeat the process for each mini-rectangle within D, and add them all up to form the Riemann sum.

$$\sum_{i,j=1}^{n} f(x_i^*, y_j^*) \Delta A_{ij}$$

The weird double counter in the summation just indicates that we cover every combination of i and j from 1 to n (we need to cover every mini-rectangle). As we let n increase, we get more and more mini-rectangles with smaller and smaller areas individually. As in the single variable case, our approximation to whatever quantity we are trying to compute with this Riemann sum will get more accurate as the number of mini-rectangles increases. We therefore define the **double integral** of $f(x, y)$ over the rectangular domain D as

$$\iint_D f(x, y)\, dA = \lim_{n \to \infty} \sum_{i,j=1}^{n} f(x_i^*, y_j^*) \Delta A_{ij}$$

So essentially, we split our two dimensional domain into a bunch of mini-rectangles, multiply the area of each mini-rectangle by the value of the integrand function at some point inside the mini-rectangle, then add all these quantities up.

A common way to interpret the double integral is the *volume* under the part of the graph of $f(x, y)$ which lies over the domain of integration D. Assuming $f(x, y)$ is always positive over D, the quantity $f(x_i^*, y_j^*) \Delta A_{ij}$ in the Riemann sum represents the volume of a thin rectangular box which stretches from the xy plane up to the height of the graph. Adding up all the volumes of the rectangular boxes will give you the total volume of the region beneath the graph of the function. This interpretation clearly parallels the area under the curve interpretation of a single integral, but again we emphasize that this is only one of many things we can do with double integrals.

Example. Evaluate $\iint_D 3\,dA$, where D is the rectangle in the xy plane given by $2 \le x \le 5$, $1 \le y \le 3$. The graph of the integrand function $f(x, y) = 3$ is just a horizontal plane situated at $z = 3$. The region below this plane and above the domain D is simply a rectangular box with height 3 and base of area 6. Since the double integral gives the volume under the graph,

$$\iint_D 3\,dA = 3 \cdot 6 = 18$$

There is another way to think about this. We can pull a constant out of the integral, so the expression becomes

$$3 \iint_D dA$$

The double integral of 1 just means we are adding up the areas of all the mini-rectangles which make up D. Therefore, the double integral $\iint_D dA$ is always just going to equal the area of D. This is 6, then we multiply by the constant outside to get the same answer of 18.

Example. Suppose we have a very thin, rectangular sheet of metal lying in the xy plane. The density in mass per area at a point (x, y) on the sheet is given by the function $\delta(x, y)$. What is the mass of the entire sheet?

We can't just multiply the density by the area of the sheet because the density varies from point to point. First, we imagine dividing up the metal sheet into many mini-rectangles, each with some area ΔA (we don't bother with all the subscript i and j business for simplicity). The mass of each mini-rectangle will be its area times the density at a point within the mini-rectangle, the assumption being that each mini-rectangle is so small that the density on that rectangle is essentially constant.

$$\Delta m = \delta(x, y)\Delta A$$

If we add up the masses of all the mini-rectangles which make up the metal sheet, we get the entire mass. This is a double integral where our domain is the metal sheet and our integrand is the density function.

$$m = \iint_D \delta(x, y)\,dA$$

Triple Integrals

The last case we will cover is an integral of a three variable function $f(x, y, z)$ over a three dimensional domain, a rectangular box in xyz space. We follow the same procedure as before to make a Riemann sum.

Suppose our rectangular box D is defined by $a \le x \le b$, $c \le y \le d$, $e \le z \le f$. First, we take a partition of x, y, and z, splitting up each axis into many subintervals. These partitions split up our box D into many mini-boxes. Let V_{ijk} be the mini-box determined by the ith subinterval of x, the jth subinterval of y, and the kth subinterval of z, and let ΔV_{ijk} denote its volume.

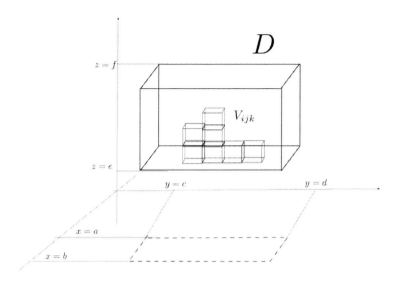

In each mini-box, we pick a random point (x_i^*, y_j^*, z_k^*) and find the value of our integrand function at that point $f(x_i^*, y_j^*, z_k^*)$. We then multiply this value by the volume of that particular mini-box, repeat the process for every mini-box in D, and add up the results.

$$\sum_{i,j,k=1}^{n} f(x_i^*, y_j^*, z_k^*)\Delta V_{ijk}$$

As we increase n in this Riemann sum, the mini-boxes will grow more numerous and smaller in size, and the approximation to whatever quantity we are trying to compute will get more accurate. So we define the ***triple integral*** of an integrand $f(x, y, z)$ over the rectangular box region D by

$$\iiint_D f(x, y, z)\, dV = \lim_{n \to \infty} \sum_{i,j,k=1}^{n} f(x_i^*, y_j^*, z_k^*)\Delta V_{ijk}$$

So we are dividing our three dimensional domain of integration into mini-boxes, then multiplying the volume of each mini-box by the value of the integrand function at some point inside the mini-box. Notice that, unlike single and double integrals, it is hard to find a simple geometrical meaning such as area under the curve or volume under the surface. The direct analogy for the triple integral would be the hypervolume of a four dimensional object, but no one wants to try to visualize that, so we do not really use this interpretation. However, there are still uses for the triple integral.

Example. Evaluate $\iiint_D 2\, dV$, where D is the rectangular box defined by $2 \le x \le 4$, $1 \le y \le 5$, $0 \le z \le 3$. We can move the constant out of the integral, so we just have to evaluate $\iiint_D dV$. This just tells us to add up the volumes of all the mini-boxes which make up D, therefore the triple integral of 1 is just the volume of our domain of integration.

$$\iiint_D 2\, dV = 2 \iiint_D dV = 2(2)(4)(3) = 48$$

Example. Suppose we have a bunch of bears squished into a rectangular box. If we know the bear density in bears per volume at a particular point is given by the function $B(x, y, z)$, how many total bears are in the box? First, imagine dividing our rectangular box into many mini-boxes, each with volume ΔV. Across each of these mini-boxes, the bear density is approximately constant, so the number of bears is the volume times the bear density.

$$\Delta \text{bears} = B(x, y, z)\Delta V$$

If we do this for each mini-box, then to get the total number of bears we just add up the number of bears in each mini-box. This is a triple integral of the bear density function over the rectangular box.

$$\text{bears} = \iiint_D B(x, y, z)\, dV$$

Exercises

1. Compute the integral

   (a) $\iint_D 5\, dA$, where D is the rectangle determined by $-1 \le x \le 3$, $0 \le y \le 2$

   (b) $\iint_D y\, dA$, where D is the square $0 \le x \le 1$, $0 \le y \le 1$

   (c) $\iiint_D -2\, dV$, where D is the rectangular box $0 \le x \le 2$, $1 \le y \le 3$, $2 \le z \le 5$

2. Consider the integral $\iint_D x + y\, dA$ where D is a square determined by $0 \le x \le 4$, $0 \le y \le 4$. Use a Riemann sum to estimate the value; divide the domain into four equal mini-squares and choose the center of each square as the point to evaluate the integrand at

Answers

1. (a) 40

   (b) $\frac{1}{2}$; the graph of $z = y$ is a diagonal plane; region under it and above the square of sides 1 is half of the cube whose edges are all 1

   (c) -24

2. 64; each mini-square has sides of length 2 and an area of 4; the centers of the squares are $(1, 1)$, $(1, 3)$, $(3, 1)$, and $(3, 3)$, so the sum should be

$$4\big[f(1, 1) + f(1, 3) + f(3, 1) + f(3, 3)\big]$$

4.2   Iterated Integrals

In the last section, we defined double and triple integrals of a function over a rectangle or rectangle box domain. But how do we actually compute these? The answer, as it turns out, is pretty simple and intuitive, although the actual proof is quite complicated so instead we will look at a somewhat convincing argument.

Volume Under a Graph

We know that if $f(x, y)$ is a positive function, then the double integral $\iint_D f(x, y)\, dA$ can be interpreted as the volume of the region under the graph of $f(x, y)$ and over the domain D in the xy plane. Let's see if we can come up with an alternate way of finding the volume under the graph which may be easier to compute.

In single variable calculus, you probably learned how to do volume by cross sections. If we had some solid which stretched from $x = a$ to $x = b$, then to find the total volume of the solid we used the integral

$$V = \int_a^b A(x)\, dx$$

where $A(x)$ gave the area of the cross section at a particular x value. This works because if you think about splitting up the x axis into many subintervals, and therefore the solid into many slices, then the volume of the slice at x_i is its cross sectional area $A(x_i)$ times its width, which is the length of the subinterval Δx_i. Adding up the volumes of each slice forms a Riemann sum, so the total volume is given by an integral.

Now imagine that this solid is embedded in three dimensional space, where its base is a rectangle region D in the xy plane defined by $a \leq x \leq b$, $c \leq y \leq d$, and its top is the graph of a function $f(x, y)$. To apply this cross sections method, we need to find an expression for $A(x)$. If we take a cross section of the solid perpendicular to the x axis at some fixed value x_i, it looks like the region under a curve in the yz plane. We know how to compute the area under a curve; it is just the single integral. So the area of the cross section at x_i is given by

$$A(x_i) = \int_c^d f(x_i, y)\, dy$$

Notice that the integrand is essentially a single variable function because x is being held constant at a value of x_i. The volume of the cross sectional slice at x_i is therefore $A(x_i)\Delta x_i = \int_c^d f(x_i, y)\, dy\, \Delta x_i$. This suggests that the total volume of the solid will be the expression

$$V = \int_a^b \left(\int_c^d f(x, y)\, dy \right) dx$$

We have an integral inside an integral! It is important to remember that in the inside integral, x is being held constant. So you integrate with respect to y while pretending that x is a constant, similar to partial differentiation.

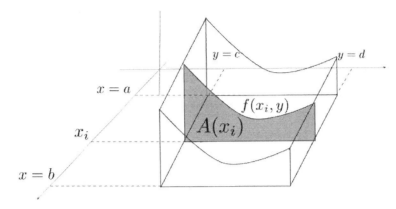

Since the volume of this region must also equal the double integral of $f(x,y)$ over D, we have two expressions for the same quantity.

$$V = \iint_D f(x,y)\, dA = \int_a^b \left(\int_c^d f(x,y)\, dy \right) dx$$

Note that we could go through the same process but take cross sections perpendicular to the y axis instead. Then we would get an expression where the inside integral is with respect to x and the outside is respect to y.

Fubini's Theorem

This result suggests that double integrals can be evaluated by computing two single variable integrals. This type of expression, with an integral inside an integral, is called an ***iterated integral***. It turns out that triple integrals can be evaluated in this way as well. These results are stated in ***Fubini's Theorem***.

Theorem. If D is a rectangle domain in the xy plane defined by $a \leq x \leq b$, $c \leq y \leq d$, then the double integral of $f(x,y)$ over D can be computed by

$$\iint_D f(x,y)\, dA = \int_a^b \int_c^d f(x,y)\, dy\, dx$$

If D is a rectangle box domain in xyz space defined by $a \leq x \leq b$, $c \leq y \leq d$, $e \leq z \leq f$, then the triple integral of $f(x,y,z)$ over D can be computed by

$$\iiint_D f(x,y,z)\, dV = \int_a^b \int_c^d \int_e^f f(x,y,z)\, dz\, dy\, dx$$

The order of integration in both cases can be switched, i.e.

$$\int_a^b \int_c^d f(x,y)\, dy\, dx = \int_c^d \int_a^b f(x,y)\, dx\, dy$$

and the same for all 6 possible orders of the triple integral.

Example. Compute the double integral $\iint_D x + y \, dA$, where D is the rectangle defined by $1 \le x \le 3$, $2 \le y \le 5$. First, we rewrite the double integral as an iterated integral. The order for rectangle domains doesn't matter, so we randomly choose to integrate with respect to y first, then x.

$$\iint_D x + y \, dA = \int_1^3 \int_2^5 x + y \, dy \, dx$$

For the inside integral, we pretend that x is a constant.

$$\int_1^3 \left[xy + \frac{1}{2}y^2 \right]_2^5 dx$$

$$\int_1^3 (5x + \frac{25}{2}) - (2x + 2) \, dx$$

$$\int_1^3 3x + \frac{21}{2} \, dx$$

Now this is just a regular single integral with respect to x.

$$\left[\frac{3}{2}x^2 + \frac{21}{2}x \right]_1^3$$

$$(\frac{27}{2} + \frac{63}{2}) - (\frac{3}{2} + \frac{21}{2}) = 45 - 12 = 33$$

Example. Compute the triple integral $\iiint_D xy^2 z \, dV$, where D is the rectangle box defined by $0 \le x \le 3$, $1 \le y \le 2$, $3 \le z \le 6$. Again, the order of integration for a rectangle box doesn't matter, so we randomly choose to integrate with respect to z first, then y, then x last.

$$\iiint_D xy^2 z \, dV = \int_0^3 \int_1^2 \int_3^6 xy^2 z \, dz \, dy \, dx$$

On the innermost integral, we pretend that both y and x are constants, then integrate with respect to z.

$$\int_0^3 \int_1^2 \left[\frac{1}{2}xy^2 z^2 \right]_3^6 dy \, dx$$

$$\int_0^3 \int_1^2 (\frac{36}{2}xy^2) - (\frac{9}{2}xy^2) \, dy \, dx$$

$$\frac{27}{2} \int_0^3 \int_1^2 xy^2 \, dy \, dx$$

since we can move constants out of the integral. Now we integrate with respect to y, pretending x is a constant

$$\frac{27}{2} \int_0^3 \left[\frac{1}{3}xy^3 \right]_1^2 dx$$

$$\frac{27}{2} \int_0^3 (\frac{8}{3}x) - (\frac{1}{3}x) \, dx$$

$$\frac{63}{2} \int_0^3 x \, dx$$

Lastly, we have a regular integral with respect to x.

$$\frac{63}{2} \left[\frac{1}{2} x^2 \right]_0^3 = \frac{63}{2} \cdot \frac{9}{2} = \frac{567}{2}$$

Non-Rectangular Domains

Now that we have a way to compute double and triple integrals, the big question is: what if we want to integrate over a domain which is not a rectangle or rectangle box? Certainly, there will be times when our domain of integration will not be so nicely squarish. However, if you remember all the way back to the last section, we only defined double and triple integrals over rectangle and rectangle box domains, respectively. How can we take an integral over a non-rectangle if the double integral is only *defined* on rectangles? We can just make the domain into a rectangle!

Suppose we want to integrate some function $f(x, y)$ over a region D in the xy plane which is the region below the line $y = x$ from $x = 0$ to $x = 2$. So our domain of integration is a triangle, which is obviously not a rectangle. However, notice that this triangle fits inside a rectangle defined by $0 \leq x \leq 2$, $0 \leq y \leq 2$. If we call this new rectangle D', then we can integrate $f(x, y)$ over D' because it is a rectangle. However, just changing the domain like that would probably change our answer too.

To fix this we define a new function $\varphi(x, y)$ which has a value of 1 when (x, y) is in our original domain D and a value of 0 otherwise. Therefore, the product $f(x, y)\varphi(x, y)$ will equal $f(x, y)$ on D and 0 at all points in the bigger rectangle D' which are not in the triangle D. It is then reasonable to assume that

$$\iint_D f(x, y) \, dA = \iint_{D'} f(x, y)\varphi(x, y) \, dA$$

When actually doing double integrals, we will not mention any of this stuff, but this is a glimpse of how integrating over non-rectangular domains is actually legal. Now we want to know how to set up an iterated integral to compute a double integral over such a domain. In general, for non-rectangular domains, changing the order of integration will change the bounds on the iterated integrals, so we must carefully choose our order.

First let's choose to integrate with respect to y first and x last. Over the entire region D, what values does x range between? The smallest value x achieves is at the origin, and the largest is at $x = 2$. So the bounds on our outer integral will be

$$\int_0^2 \int_{something}^{something} f(x, y) \, dy \, dx$$

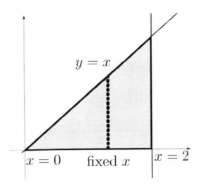

To determine the bounds on the inner integral, we ask ourselves: for a *fixed* value of x, what values does y range between? You can imagine picking any value of x between 0 and 2, drawing vertical line, and asking at what values of y does the line cross our domain D. For any fixed x value, y starts at 0 and goes until it hits the line $y = x$. So the bounds for y will be $0 \le y \le x$, and therefore our iterated integral is

$$\iint_D f(x,y)\, dA = \int_0^2 \int_0^x f(x,y)\, dy\, dx$$

If we wanted to do the opposite order, x first and y last, to determine the outer bounds we would ask: over the *entire* region, what values does y range between? The answer is $0 \le y \le 2$, so those are the bounds on the outer integral. For the inner integral, we ask: for a *fixed* value of y, what values does x range between? In other words, if we draw a horizontal line at a fixed y, at what values of x will the line intersect our triangle domain? x starts at the line $y = x$ and goes right until it hits $x = 2$, so the bounds will be $y \le x \le 2$ and the iterated integral becomes

$$\iint_D f(x,y)\, dA = \int_0^2 \int_y^2 f(x,y)\, dx\, dy$$

Finding the bounds for double and triple integrals is usually the hardest part, since actually computing them is just a combination of single integrals. Over the next few sections, we will go through many examples of finding the right boundaries for iterated integrals.

Exercises

1. Compute the double or triple integral

   (a) $\iint_D x + y\, dA$, where D is the rectangle $0 \le x \le 2$, $0 \le y \le 2$

   (b) $\iiint_D x^2 y + 3z\, dV$, where D is the rectangle box $0 \le x \le 3$, $-1 \le y \le 1$, $0 \le z \le 2$

   (c) $\iint_D x + y^2\, dA$, where D is the region under the line $y = x$ from $0 \le x \le 2$

   (d) $\iint_D e^{-x^2}\, dA$, where D is the region under the line $y = 2x$ from $0 \le x \le 2$

Answers

1. (a) 8

   (b) 36

   (c) 4; one possible way:

$$\int_0^2 \int_0^x x + y^2 \, dy \, dx = \int_0^2 \left[xy + \frac{1}{3}y^3 \right]_0^x dx$$

$$= \int_0^2 x^2 + \frac{1}{3}x^3 \, dx$$

$$= \left[\frac{1}{3}x^3 + \frac{1}{12}x^4 \right]_0^2 = \frac{8}{3} + \frac{4}{3}$$

   (d) $-e^{-4} + 1$; setup:

$$\int_0^2 \int_0^{2x} e^{-x^2} \, dy \, dx$$

4.3   Setting Up Integrals

In this section we will go through many examples of finding bounds of integration for iterated integrals. It is hard to come up with specific rules to follow every time, so you will have to get a feeling for it by doing lots of practice.

One thing to note is that the integrand function has no impact on the bounds of integration. In fact, we will completely ignore the integrand function for most of the problems in this section. Our main goal is to visualize the domain of integration and identify the bounds for the iterated integrals. Once we have those, the computation is straightforward because they are just single integrals.

Double Integrals

The general procedure for finding bounds on two dimensional domains is:

1. Choose which order you will integrate in, either x first or y first. You should choose whichever order is easier based on the domain given. For example, if the curves in the domain are expressed in the form $y = f(x)$, then it might be easier to do y first.

2. To find the bounds on the outer iterated integral, look at the variable you chose to integrate *second* and ask: over the *entire* region, what values does the variable range over?

3. For the inner integral, ask: for a *fixed* value of the second variable, what values does the first variable range over? The bounds for the first variable will often be expressions involving the second.

Example. Write $\iint_D f(x,y)\, dA$ as an iterated integral, where D is the region in the xy plane bounded by the parabolas $y = x^2$ and $y = 8 - x^2$. Both curves which determine our domain are expressed as functions of y in terms of x, so it is much more convenient to integrate with respect to y first and x second.

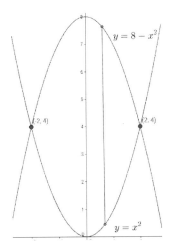

To determine what values x ranges over across the entire region, we need to find the intersection of the parabolas. We find that they intersect where $x = -2$ and $x = 2$, so our outer bounds are

$$\int_{-2}^{2} \int f(x,y)\, dy\, dx$$

For a fixed value of x, we see that y ranges from the bottom parabola to the top parabola. On the bottom is $y = x^2$ and the top is $y = 8 - x^2$. Another way to think about it is: if I draw a vertical line at a fixed value of x, where does it intersect our domain? Again, we see that this line will intersect D at the bottom and top parabolas.

$$\iint_D f(x,y)\, dA = \int_{-2}^{2} \int_{x^2}^{8-x^2} f(x,y)\, dy\, dx$$

Example. Write an iterated integral where D is the unit circle $x^2 + y^2 = 1$. For a circle, the order of integration hardly changes anything, so we will do x first just for some variety.

Over the entire circle, y ranges from -1 to 1. For a fixed value of y, x goes from the left side of the circle to the right side. The left and right sides of the unit circle can be written as functions of x in terms of y, which we obtain by solving for x in $x^2 + y^2 = 1$.

$$\iint_D f(x,y)\, dA = \int_{-1}^{1} \int_{-\sqrt{1-y^2}}^{\sqrt{1-y^2}} f(x,y)\, dx\, dy$$

You can check that if we choose the order y, x, we would have gotten

$$\iint_D f(x,y)\, dA = \int_{-1}^{1} \int_{-\sqrt{1-x^2}}^{\sqrt{1-x^2}} f(x,y)\, dy\, dx$$

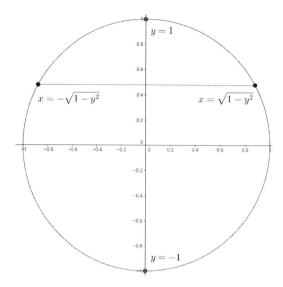

Triple Integrals

The general procedure for finding bounds on three dimensional domains is:

1. Choose the order of integration based on the bounds given. For example, if the region is bounded by surfaces where z is given as a function of x and y, then integrating with respect to z first makes sense.

2. Ignore the first variable for now, and visualize the projection of the region onto the plane determined by the second and third variables. For example, if you made z first, then look at the projection of the region onto the xy plane.

3. Determine the bounds on the second and third variables as if it were a double integral on this projection.

4. Determine the bounds on the first variable by asking: for a *fixed* value of the second and third variables, what values does the first variable range between?

Example. Write $\iiint_D f(x, y, z)\, dV$ as an interated integral, where D is the unit ball (the unit sphere filled up). As with a circle in two dimensions, the order for a ball hardly matters. We will choose the order z, y, x. The projection of the ball onto the xy plane is just the unit circle. One way to visualize the projection of a solid onto a particular plane is to imagine the solid being squished onto that plane.

To find the bounds for y and x, we act like we are doing a double integral over the unit circle. We did this in the last example, and the bounds for the order y, x are

$$\int_{-1}^{1} \int_{-\sqrt{1-x^2}}^{\sqrt{1-x^2}} \int f(x, y, z)\, dz\, dy\, dx$$

For the bounds on the z integral, we ask ourselves what values z ranges over for a fixed x and y. Imagine picking a random point (x, y) within the unit circle and drawing a vertical

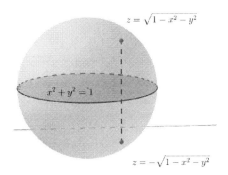

line perpendicular to the z axis. This line would intersect our domain at the bottom and top of the sphere. The bottom and upper hemispheres can be expressed as functions of z in terms of x and y by solving for z in the equation $x^2 + y^2 + z^2 = 1$. Therefore, our iterated integral is

$$\iiint_D f(x,y,z)\, dV = \int_{-1}^{1} \int_{-\sqrt{1-x^2}}^{\sqrt{1-x^2}} \int_{-\sqrt{1-x^2-y^2}}^{\sqrt{1-x^2-y^2}} f(x,y,z)\, dz\, dy\, dx$$

Example. Write an iterated integral where D is the solid bounded by the planes $z = 0$, $y = x$, $x + y + z = 4$, and $x = 0$. This shape is a tetrahedron (shape with four triangle faces).

We choose the order z, y, x because we have $z = 0$ and $x + y + z = 4$, which can be easily solved for z in terms of the other two variables. y would have been another fine choice for the first variable, but x would be troublesome as we would have to split the integral into two parts.

The projection of this region onto the xy plane is a triangle defined by the y axis and the lines $y = x$ and $y = -x + 4$. The second line was obtained by setting $z = 0$ on the plane $x + y + z = 4$, which finds its intersection with the xy plane.

Now we pretend we are finding a double integral over this triangle. Over the whole region, x ranges from 0 to 2, the point where the two lines intersect. For a fixed x, y ranges from the bottom line $y = x$ to the top line $y = -x + 4$.

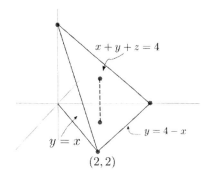

For a fixed point (x, y), z goes from $z = 0$ up until it hits the plane $x + y + z = 4$. Therefore, our iterated integral is

$$\iiint_D f(x, y, z)\, dV = \int_0^2 \int_x^{-x+4} \int_0^{4-x-y} f(x, y, z)\, dz\, dy\, dx$$

Exercises

1. Given a domain, find bounds for the iterated integral

    (a) The first bump of $y = \sin x$

    (b) The region bounded by $y = x^2$, $y = 4$, and $y = -3x + 4$ (you will need two integrals)

    (c) The solid bounded by $z = x^2 + y^2$ and $z = 2 - x^2 - y^2$

Answers

1. (a) $\int_0^\pi \int_0^{\sin x} dy\, dx$

    (b) $\int_{-2}^0 \int_{x^2}^4 dy\, dx + \int_0^1 \int_{x^2}^{-3x+4} dy\, dx$

    (c) $\int_{-1}^1 \int_{-\sqrt{1-x^2}}^{\sqrt{1-x^2}} \int_{x^2+y^2}^{2-x^2+y^2} dz\, dy\, dx$; the two paraboloids intersect at $z = 1$; the projection onto the xy plane is the unit circle

4.4   Change of Variables

Just like we made u-substitutions or trig substitutions to evaluate single integrals, we often want to use substitutions to simplify multiple integrals. Substitutions in single variable calculus were made to simplify the integrand into a function that we knew the antiderivative of. However, we will see that in multiple integrals substitutions are often made to simplify the domain of integration as well as the integrand.

As an example, say our domain of integration R is the part of the unit circle in the first quadrant. Our bounds of integration would look something like this:

$$\iint_R f(x,y)\,dA = \int_0^1 \int_0^{\sqrt{1-x^2}} f(x,y)\,dy\,dx$$

These bounds look pretty ugly, so we might want to find an appropriate substitution to simplify the domain. What coordinate system do we know where equations of circles become very simple? Polar coordinates!

Consider the function $T(r,\theta) = \langle r\cos\theta, r\sin\theta \rangle$, which you can call the polar coordinates transformation. It takes a point in the imaginary $r\theta$ plane and converts it to a point in the xy plane. If we used polar coordinates to describe our circle, then r simply ranges from 0 to 1, and θ from 0 to $\frac{\pi}{2}$. These bounds are all constants, so this suggests that in the $r\theta$ plane, the region is just a rectangle D defined by $0 \le r \le 1$, $0 \le \theta \le \frac{\pi}{2}$.

The quarter circle in the xy plane R is the *image* of the rectangle in the $r\theta$ plane under the transformation $T(r,\theta) = \langle r\cos\theta, r\sin\theta \rangle$, so we can say $R = T(D)$. In other words, if we took every point in the rectangle $0 \le r \le 1$, $0 \le \theta \le \frac{\pi}{2}$ and fed them into the function $T(r,\theta)$, then our new region would be a quarter circle in the xy plane.

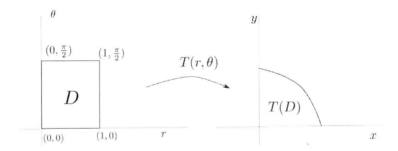

If we can somehow transform our integral over $T(D)$ into an integral over D, then the iterated integrals would be much easier to evaluate since the bounds are constant.

$$\iint_{T(D)} f(x,y)\,dA = \iint_D f(T(r,\theta))(????)\,dA$$

The question marks signify that we must multiply our integrand by some factor in order to transform the integral, just like in u-substitution. Also notice that the integrand function is now in terms of r and θ instead of x and y. In other words, in addition to multiplying by some factor we must also express all the x's and y's in terms of r and θ, similar to how in u-sub we rewrite x in terms of u.

The Formula

The general problem is this: we have an integral over a domain R in the xy plane, and this region is the image of a simpler region in the uv plane under the transformation

$T(u,v)$. We want to transform our integral over R in xy, which can write as $T(D)$, into an integral over D in uv. Note that u and v just denote generic variables; in the above discussion r and θ played the roles of u and v.

Theorem. Let $T : \mathbb{R}^2 \mapsto \mathbb{R}^2$ be a function which transforms uv coordinates to xy coordinates. If we are trying to integrate $f(x,y)$ over a region in the xy plane which is the image of the region D in the uv plane under the transformation $T(u,v)$, then

$$\iint_{T(D)} f(x,y)\, dA = \iint_D f(T(u,v))\, |\det T'|\ dA$$

So to transform our integral into an integral over the simpler region in the uv plane, we must do two things to the integrand: rewrite all the x's and y's in terms of u and v, and multiply by the *absolute value* of the *determinant* of the Jacobian matrix of T.

Here's another way to think about it. Imagine dividing up our rectangle D in the uv plane into many mini-rectangles, each with an area of ΔA. When each mini-rectangle is transformed into the xy plane, it gets twisted and deformed by T, so its area is no longer ΔA. We need to find out how much the area of each mini-rectangle changes, and this "stretching factor" turns out to be the absolute value of the Jacobian determinant $|\det T'|$. So the area of each mini-patch in the xy plane corresponds to the area of a mini-rectangle in the xy plane multiplied by this stretching factor. In symbols,

$$\Delta A_{xy} = |\det T'|\, \Delta A_{uv}$$

The change of variables formula basically says to substitute this expression into the original integral to replace dA.

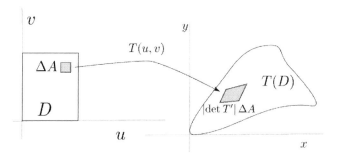

As a first example, let's go back to the problem we considered at the beginning of the section.

Example. We want to compute a double integral $\iint_R x+y\, dA$, where R is the part of the unit circle in the first quadrant. We have seen that this region is the image of a rectangle D in the $r\theta$ plane under the transformation $T(r,\theta) = \langle r\cos\theta, r\sin\theta \rangle$.

To transform the integral into the $r\theta$ plane, we need to compute the absolute value of the determinant of the Jacobian of this transformation.

$$T'(r, \theta) = \begin{bmatrix} \cos\theta & -r\sin\theta \\ \sin\theta & r\cos\theta \end{bmatrix}$$

$$\det T' = r\cos^2\theta + r\sin^2\theta = r$$

If we restrict r to positive values, then we don't have to bother with the absolute value. So we have to multiply the integrand by this factor r, as well as replace all the x's and y's by $r\cos\theta$ and $r\sin\theta$, respectively.

$$\iint_R x + y \, dA = \iint_D (r\cos\theta + r\sin\theta)r \, dA$$

We can then write our new integral as an iterated integral, since we know D is the rectangle defined by $0 \le r \le 1$, $0 \le \theta \le \frac{\pi}{2}$

$$\int_0^{\pi/2} \int_0^1 r^2(\cos\theta + \sin\theta) \, dr \, d\theta$$

We will look at integrals using the polar coordinates transformation more in the next section. We do not have to have a special coordinate system, such as polars or sphericals, in mind when making a change of variables. The transformation function is often naturally motivated by a particularly odd expression in the integrand or the domain.

Example. Use a double integral to find the area of an ellipse

$$\frac{x^2}{a^2} + \frac{y^2}{b^2} = 1$$

Remember that the double integral of 1 will just find the area of the domain of integration, so we need to compute

$$\iint_R dA$$

where R is the above ellipse. The problem is that finding the bounds of integration will be a big pain, so let's see if we can find a suitable transformation function. Notice if we define

$$u = \frac{x}{a} \qquad \text{and} \qquad v = \frac{y}{b}$$

then the ellipse equation becomes $u^2 + v^2 = 1$. So if we make the above substitutions, then the corresponding region in the uv plane will just be the unit circle! This motivates the transformation function

$$T(u, v) = \langle au, bv \rangle$$

All we have to do now is calculate the Jacobian determinant of this thing

$$T'(u, v) = \begin{bmatrix} a & 0 \\ 0 & b \end{bmatrix}$$

$$\det T' = ab$$

Therefore, our transformed integral is

$$\iint_R dA = \iint_D ab\, dA$$

where D is the unit circle in the uv plane. Since a and b are constants, we can move them out of the integral, which then just becomes the integral of 1 over the unit circle. Therefore, the area of the ellipse is

$$ab \iint_D dA = \pi ab$$

Example. Say we want to compute the double integral $\iint_R x + y\, dA$, where R is the image of a square D in the uv plane $0 \le u \le 1$, $0 \le v \le 1$. The transformation between uv coordinates and xy coordinates is given by $T(u,v) = \langle u^2 + 4v, v^3 + 2u \rangle$. Use change of variables to convert the integral over R to a new integral over D in the uv plane.

We have no idea what our domain in the xy plane R looks like, but all we need to know is the transformation function T and what the domain looks like in the uv plane. In this case, we are told that it's a square, so we know that an iterated integral over D will be very simple. Really, the only thing we have to do is compute the determinant of the Jacobian; once we have that we just have to plug stuff in.

$$T'(u,v) = \begin{bmatrix} 2u & 4 \\ 2 & 3v^2 \end{bmatrix}$$

$$\det T'(u,v) = \begin{vmatrix} 2u & 4 \\ 2 & 3v^2 \end{vmatrix}$$
$$= 6uv^2 - 8$$

Since u and v are always less than 1, the determinant will always be negative. Therefore, its absolute value is $8 - 6uv^2$. Now we can transform our integral by multiplying by this factor and expressing $x + y$ in terms of u and v.

$$\iint_R x + y\, dA = \iint_D \left((u^2 + 4v) + (v^3 + 2u)\right)(8 - 6uv^2)\, dA$$

Since D is just a square in the uv plane, we can write an iterated integral

$$\int_0^1 \int_0^1 \left((u^2 + 4v) + (v^3 + 2u)\right)(8 - 6uv^2)\, du\, dv$$

Example. Calculate $\iint_R x^2 + y^2\, dA$, where R is the region bounded by the hyperbolas $xy = 2$, $xy = 4$, $x^2 - y^2 = 1$, and $x^2 - y^2 = 3$. If we choose new variables $u = xy$ and

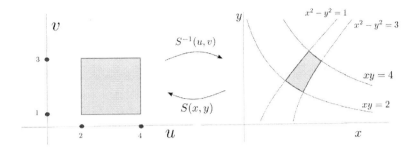

$v = x^2 - y^2$, then over our region R, u ranges from 2 to 4 and v ranges from 1 to 3, so in the uv plane R becomes a simple rectangle. We can write this transformation between coordinates as $S(x, y) = \langle xy, x^2 - y^2 \rangle$.

Notice that the function S converts x and y into u and v, the opposite of the transformation that we want. In fact, it is the *inverse* of $T(u, v)$ which transforms u and v into x and y, which is what we need for the change of variables. Remember that we want to find out how much a small area in the uv plane is stretched when it goes to the xy plane.

$$\Delta A_{xy} = (\text{stretching factor}) \, \Delta A_{uv}$$

Since $S(x, y)$ maps from xy to uv coordinates, we can think of it as the stretching factor for the opposite process: when a small area in the xy plane is transformed to the uv plane, it is stretched according to this function S.

$$\Delta A_{uv} = |\det S'| \, \Delta A_{xy}$$

Therefore, the stretching factor that we need to put in our integral is

$$\Delta A_{xy} = \frac{1}{|\det S'|} \Delta A_{uv}$$

Let's compute that now:

$$\det S' = \begin{vmatrix} y & x \\ 2x & -2y \end{vmatrix}$$
$$= -2y^2 - 2x^2 = -2(x^2 + y^2)$$
$$\frac{1}{|\det S'|} = \frac{1}{2(x^2 + y^2)}$$

The absolute value doesn't matter because this expression is always positive. We still need to get this in terms of u and v, but we will automatically get that when we find our integrand in terms of u and v, since our integrand is $x^2 + y^2$. To do this, you just have to kind of play around.

$$u^2 = x^2 y^2 \qquad \text{and} \qquad v^2 = x^4 - 2x^2 y^2 + y^4$$

$$4u^2 + v^2 = x^4 + 2x^2y^2 + y^4 = (x^2 + y^2)^2$$

$$\text{Therefore, } x^2 + y^2 = \sqrt{4u^2 + v^2}$$

Now we can finally transform our integral

$$\iint_R x^2 + y^2 \, dA = \iint_D \sqrt{4u^2 + v^2} \cdot \frac{1}{2\sqrt{4u^2 + v^2}} \, dA$$
$$= \frac{1}{2} \iint_D dA$$
$$= \frac{1}{2} \int_2^4 \int_1^3 dv \, du = 2$$

Note that we didn't actually need to find $x^2 + y^2$ in terms of u and v because it cancels out with the stretching factor anyway, but that will not always be the case, so we showed how to do it.

Proof

We will now attempt to give a mildly convincing argument of the change of variables formula for two variables. Again, we want to calculate the "stretching factor" that a small area in the uv plane gets multiplied by when it gets transformed into the xy plane by the function $T(u, v)$.

$$\Delta A_{xy} = (\text{stretching factor}) \, \Delta A_{uv}$$

Let's look at a small square in the uv plane whose sides have lengths Δu and Δv. The area of the square is then $\Delta A_{uv} = \Delta u \Delta v$. We will denote the bottom left corner of the square by the point (u_0, v_0), so the bottom right and top left corners become $(u_0 + \Delta u, v_0)$ and $(u_0, v_0 + \Delta v)$, respectively.

We are interested in the image of this square on the xy plane under the transformation $T(u, v)$. The key idea here is that the square will be transformed into a shape in the xy plane which is approximately a parallelogram, whose area is ΔA_{xy}.

We know that the area of a parallelogram can be computed as the magnitude of the cross product of two vectors which make up the sides of it, so we need to represent the sides of this parallelogram as vectors.

As you can see in the picture, the bottom edge of the parallelogram has endpoints $T(u_0, v_0)$ and $T(u_0 + \Delta u, v_0)$ which correspond to the endpoints of the bottom edge of the square in the uv plane. We can then define the vector

$$T(u_0 + \Delta u, v_0) - T(u_0, v_0)$$

which will point along this edge.

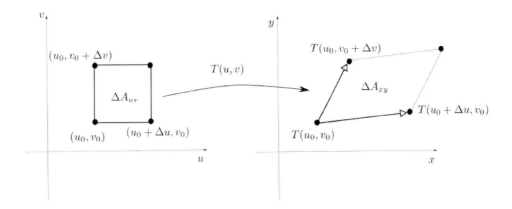

Remember that both the square in the uv plane and the parallelogram in the xy plane are supposed to be very small in size. Therefore the displacement Δu is small enough that we can approximate the above vector with the *differential* of T at the point (u_0, v_0) with displacement vector $\langle \Delta u, 0 \rangle$.

$$T(u_0 + \Delta u, v_0) - T(u_0, v_0) \approx dT_{(u_0, v_0)}(\Delta u \, \mathbf{e}_1)$$

$$= \begin{bmatrix} \dfrac{\partial T_1}{\partial u} & \dfrac{\partial T_1}{\partial v} \\[2mm] \dfrac{\partial T_2}{\partial u} & \dfrac{\partial T_2}{\partial v} \end{bmatrix} \begin{bmatrix} \Delta u \\[2mm] 0 \end{bmatrix}$$

$$= \frac{\partial T}{\partial u} \Delta u$$

Keep in mind that since $T(u, v)$ is a vector valued function, its partial derivatives are vectors. We can repeat this process to find a vector which points along the left edge of the parallelogram.

$$T(u_0, v_0 + \Delta v) - T(u_0, v_0) \approx dT_{(u_0, v_0)}(\Delta v \, \mathbf{e}_2)$$

$$= \begin{bmatrix} \dfrac{\partial T_1}{\partial u} & \dfrac{\partial T_1}{\partial v} \\[2mm] \dfrac{\partial T_2}{\partial u} & \dfrac{\partial T_2}{\partial v} \end{bmatrix} \begin{bmatrix} 0 \\[2mm] \Delta v \end{bmatrix}$$

$$= \frac{\partial T}{\partial v} \Delta v$$

The area of the parallelogram will then be given by the magnitude of the cross product of these vectors.

$$\Delta A_{xy} = \left| \frac{\partial T}{\partial u} \times \frac{\partial T}{\partial v} \right| \Delta u \Delta v$$

Since these are two dimensional vectors, we have to do the trick to extend them to three dimensions by adding a 0 as the third component. The end result is what we expect

$$\Delta A_{xy} = \begin{vmatrix} \dfrac{\partial T_1}{\partial u} & \dfrac{\partial T_2}{\partial u} \\[2ex] \dfrac{\partial T_1}{\partial v} & \dfrac{\partial T_2}{\partial v} \end{vmatrix} \Delta A_{uv} = \begin{vmatrix} \dfrac{\partial T_1}{\partial u} & \dfrac{\partial T_1}{\partial v} \\[2ex] \dfrac{\partial T_2}{\partial u} & \dfrac{\partial T_2}{\partial v} \end{vmatrix} \Delta A_{uv} = |\det T'| \, \Delta A_{uv}$$

Exercises

1. Find the determinant of the Jacobian of the transformation

$$T(u, v) = \langle u^2 v^3 - 2u, v^5 + 2uv \rangle$$

2. Compute the integral $\iint_R x + y \, dA$, where R is the image of the square $0 \le u \le 1$, $0 \le v \le 1$ in the uv plane under the transformation

$$T(u, v) = \langle 2uv, v^2 \rangle$$

3. Compute the integral $\iint_R x^2 y^3 \, dx \, dy$, where R is the region bounded by the curves $xy = 6$, $xy = 9$, $xy^2 = 5$, $xy^2 = 7$

Answers

1. $|10uv^7 - 2u^2 v^3 - 10v^4 - 4u|$

2. $\frac{9}{5}$

$$\iint_R x + y \, dA = \iint_D (2uv + v^2) 4v^2 \, dA = \int_0^1 \int_0^1 8uv^3 + 4v^4 \, du \, dv$$

3. 45; use the transformation function $S(x, y) = \langle xy, xy^2 \rangle$

$$\iint_R x^2 y^3 \, dA = \iint_D uv \frac{1}{v} \, dA = \int_5^7 \int_6^9 u \, du \, dv$$

4.5   Integrals in Different Coordinate Systems

In this section we will look at the most common application of the change of variables formula: evaluating double and triple integrals with polar, cylindrical, and spherical coordinates. Many types of regions such as circles, spheres, cones, and pretty much anything curvy are much more convenient to find bounds for in these coordinate systems.

Polar Coordinates

The first case we examine is using polar coordinates to calculate double integrals. We actually did this in the first example of last section, so we will just do a quick review of the process. The polar coordinates transformation function which converts r and θ to x and y is $T(r, \theta) = \langle r\cos\theta, r\sin\theta \rangle$. The stretching factor, which is the absolute value of the Jacobian determinant is

$$T'(r,\theta) = \begin{bmatrix} \cos\theta & -r\sin\theta \\ \\ \sin\theta & r\cos\theta \end{bmatrix}$$

$$\det T' = r\cos^2\theta - (-r\sin^2\theta) = r$$

We will usually have r be positive, so we don't need to worry about the absolute value. Therefore, to convert an integral to polar coordinates we just plop an r in the integrand and express all x and y in terms of r and θ.

$$\iint_{T(D)} f(x,y)\, dA = \iint_D f(r\cos\theta, r\sin\theta) r\, dA$$

Example. Evaluate $\iint_R x + y\, dA$, where R is the circle of radius 2 centered at the origin in the xy plane. This region is a simple rectangle in the $r\theta$ plane because r ranges from 0 to 2 and θ ranges from 0 to 2π.

$$\begin{aligned}
\iint_R x + y\, dA &= \iint_D (r\cos\theta + r\sin\theta) r\, dA \\
&= \int_0^{2\pi} \int_0^2 r^2(\cos\theta + \sin\theta)\, dr\, d\theta \\
&= \frac{1}{3}\int_0^{2\pi} \left[r^3(\cos\theta + \sin\theta) \right]_0^2 d\theta \\
&= \frac{8}{3}\int_0^{2\pi} \cos\theta + \sin\theta\, d\theta \\
&= \frac{8}{3}\left[\sin\theta - \cos\theta \right]_0^{2\pi} = 0
\end{aligned}$$

Of course, the regions do not always have to be rectangles in the $r\theta$ plane, as we see in the next example. Note that the most common order for polar double integrals is r then θ because many polar curves are given as functions of θ.

Example. Use a double integral to find the area of one petal of the rose $r = 2\sin(2\theta)$. Remember, to find the area of a two dimensional region we can calculate the double integral of 1 over that domain.

$$\iint_R dA = \iint_D r\, dA$$

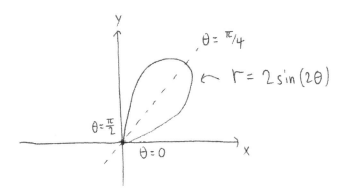

Now we just need to find bounds on r and θ so we can set up an iterated integral. r stretches out from the origin where $r = 0$ until it hits the curve, which we know is given by $r = 2\sin(2\theta)$. Over one petal, in particular the petal in the first quadrant, θ ranges from 0 to $\pi/2$.

$$
\iint_D r\, dA = \int_0^{\pi/2} \int_0^{2\sin(2\theta)} r\, dr\, d\theta
$$
$$
= 2 \int_0^{\pi/2} \sin^2(2\theta)\, d\theta
$$
$$
= 2 \int_0^{\pi/2} \frac{1}{2} - \frac{1}{2}\cos(4\theta)\, d\theta \quad \text{(double cosine identity: } \cos(2\theta) = 1 - 2\sin^2\theta)
$$
$$
= 2 \left[\frac{1}{2}\theta - \frac{1}{8}\sin(4\theta) \right]_0^{\pi/2} = \frac{\pi}{2}
$$

Cylindrical Coordinates

Although we didn't really mention triple integrals last section, the change of variables formula applies to integrals in any dimension. For a triple integral, the only difference from doubles is that the Jacobian determinant will be 3×3.

The function associated with cylindrical coordinates, which maps $r\theta z$ space to xyz space, is

$$
T(r, \theta, z) = \langle r\cos\theta, r\sin\theta, z \rangle
$$

The stretching factor is the same as that for polar coordinates, as you can see here:

$$
\det T' = \begin{vmatrix} \cos\theta & -r\sin\theta & 0 \\ \sin\theta & r\cos\theta & 0 \\ 0 & 0 & 1 \end{vmatrix}
$$

$$
= \cos\theta \begin{vmatrix} r\cos\theta & 0 \\ 0 & 1 \end{vmatrix} + r\sin\theta \begin{vmatrix} \sin\theta & 0 \\ 0 & 1 \end{vmatrix} + 0
$$

$$
= r\cos^2\theta + r\sin^2\theta = r
$$

Therefore, to convert an integral to cylindrical coordinates, we do

$$
\iiint_{T(D)} f(x,y,z)\, dV = \iiint_D f(r\cos\theta, r\sin\theta, z)r\, dV
$$

For the rest of the section, we will focus on just setting up the integral, since the actual computation is still just evaluated single integrals. The most convenient order for cylindrical coordinates is usually z first, then r, then θ.

If you choose this order, then the general strategy is similar to the one we used for triple integrals in rectangular coordinates. To determine the bounds of the outer two integrals with respect to r and θ, look at the projection of the region onto the xy plane and pretend you are setting up a double integral over this projection. The bounds on z will usually come naturally if you look at what surfaces the region is sandwiched between.

Example. Set up an iterated integral for $\iiint_R xz\, dV$, where R is the region above the paraboloid $z = x^2 + y^2$ and below the plane $z = 4$.

First, we rewrite our boundary surfaces in cylindrical coordinates, obtaining $z = r^2$ for the paraboloid. From this we can easily tell the bounds on our z integral: the region is sandwiched between $z = r^2$ and $z = 4$

For r and θ, we look at the projection of the region onto the xy plane. Again, imagine the region being squished down onto the plane. The projection is the circle $r = 2$, which you can get by plugging $z = 4$ into the paraboloid equation. From there its the same as setting up a double integral in polar coordinates on the circle.

$$
\iiint_R xz\, dV = \iiint_D (r\cos\theta z)r\, dV
$$

$$
= \int_0^{2\pi} \int_0^2 \int_{r^2}^4 r^2 \cos\theta z\, dz\, dr\, d\theta
$$

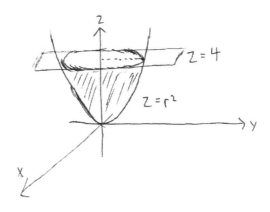

Spherical Coordinates

Lastly, we look at how to set up integrals in spherical coordinates. The transformation function is

$$T(\rho, \theta, \phi) = \langle \rho \sin \phi \cos \theta, \rho \sin \phi \cos \theta, \rho \cos \phi \rangle$$

This transforms a region in three dimensional $\rho\theta\phi$ space to a region in xyz space.

Calculating the stretching factor is quite messy, but here it is:

$$\det T' = \begin{vmatrix} \sin\phi\cos\theta & -\rho\sin\phi\sin\theta & \rho\cos\phi\cos\theta \\ \sin\phi\sin\theta & \rho\sin\phi\cos\theta & \rho\cos\phi\sin\theta \\ \cos\phi & 0 & -\rho\sin\phi \end{vmatrix}$$

$$= \sin\phi\cos\theta \begin{vmatrix} \rho\sin\phi\cos\theta & \rho\cos\phi\sin\theta \\ 0 & -\rho\sin\theta \end{vmatrix} + \rho\sin\phi\sin\theta \begin{vmatrix} \sin\phi\sin\theta & \rho\cos\phi\sin\theta \\ \cos\phi & -\rho\sin\phi \end{vmatrix}$$

$$+ \rho\cos\phi\cos\theta \begin{vmatrix} \sin\phi\sin\theta & \rho\sin\phi\cos\theta \\ \cos\phi & 0 \end{vmatrix}$$

$$= \sin\phi\cos\theta(-\rho^2\sin^2\phi\cos\theta) + \rho\sin\phi\sin\theta(-\rho\sin^2\phi\sin\theta - \rho\cos^2\phi\sin\theta)$$

$$+ \rho\cos\phi\cos\theta(-\rho\sin\phi\cos\phi\cos\theta)$$

$$= -\rho^2\sin^3\phi\cos^2\theta - \rho^2\sin\phi\sin^2\theta - \rho^2\sin\phi\cos^\phi\cos^2\theta$$

$$= -\rho^2\sin\phi(\sin^2\phi\cos^2\theta + \sin^2\theta + \cos^2\phi\cos^2\theta)$$

$$= -\rho^2\sin\phi(\cos^2\theta + \sin^2\theta)$$

$$= -\rho^2\sin\phi$$

Taking the absolute value of this determinant gets rid of the negative sign, so to convert an integral to spherical coordinates we do

$$\iiint_{T(D)} f(x,y,z)\,dV = \iiint_D f(\rho\sin\phi\cos\theta, \rho\sin\phi\sin\theta, \rho\cos\phi)\rho^2\sin\phi\,dV$$

The strategy for spherical triple integrals is a little different. Almost all of the time, the bounds on θ and ϕ will be constants, so to determine them you just have to ask what values they range between over the *entire* region. For ρ, imagine a ray coming out of the origin at a *fixed* θ and ϕ, then see where this ray intersects the region. The first variable of integration will almost always be ρ.

Example. Use a triple integral to find the volume of a sphere of radius a. For a sphere, the simplest region possible in spherical coordinates, all three bounds are constants, i.e. the region is a rectangle box in $\rho\theta\phi$ space.

$$\begin{aligned}
\iiint_R dV &= \iiint_D \rho^2\sin\phi\,dV \\
&= \int_0^{2\pi}\int_0^{\pi}\int_0^a \rho^2\sin\phi\,d\rho\,d\phi\,d\theta \\
&= \frac{1}{3}a^3\int_0^{2\pi}\int_0^{\pi}\sin\phi\,d\phi\,d\theta \\
&= \frac{2}{3}a^3\int_0^{2\pi} d\theta \\
&= \frac{4\pi}{3}a^3
\end{aligned}$$

Example. Set up an iterated integral to evaluate $\iiint_R xy\,dV$, where R is the region above the cone $z = \sqrt{x^2+y^2}$ and below the sphere $x^2+y^2+z^2 = 4$. This shape is an ice cream cone.

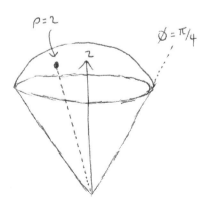

We can tell pretty clearly that ρ ranges from 0 to 2 and θ ranges from 0 to 2π everywhere, so we just need to find the bounds on ϕ. We can find this by expressing our cone in terms of spherical coordinates

$$z = \sqrt{x^2 + y^2}$$

$$\rho \cos \phi = \sqrt{\rho^2 \sin^2 \phi \cos^2 \theta + \rho^2 \sin^2 \phi \sin^2 \theta}$$

$$\rho \cos \phi = \rho \sin \phi$$

$$\tan \phi = 1$$

$$\phi = \frac{\pi}{4}$$

Therefore, our iterated integral is

$$\iiint_R xy \, dV = \iiint_D (\rho \sin \phi \cos \theta)(\rho \sin \phi \sin \theta)(\rho^2 \sin \phi) \, dV$$

$$= \int_0^{2\pi} \int_0^{\pi/4} \int_0^2 \rho^4 \sin^3 \phi \sin \theta \cos \theta \, d\rho \, d\phi \, d\theta$$

Exercises

1. Given the domain, set up the bounds on an iterated integral using either polar, cylindrical, or spherical coordinates

   (a) The region above the y axis inside the circle of radius 4 but outside the circle of radius 2

   (b) The solid above the cone $z^2 = x^2 + y^2$ and below the plane $z = 4$

   (c) The solid above the xy plane inside the sphere of radius 5 but outside the sphere of radius 2

Answers

1. (a) $\int_0^\pi \int_2^4 \, dr \, d\theta$

   (b) $\int_0^{2\pi} \int_0^4 \int_r^4 \, dz \, dr \, d\theta$

   (c) $\int_0^{2\pi} \int_0^{\pi/2} \int_2^5 \, d\rho \, d\phi \, d\theta$

Chapter 5

Vector Calculus

5.1   Line Integrals

In this section we will continue our discussion of curves in two and three dimensional space which we started back in section 3.1. First of all, what is a curve? A curve is just a set of points, which we will call C, that is the image of some single variable vector function. For example, the image of the function

$$\gamma(t) = \langle \cos t, \sin t \rangle$$

is the unit circle in two dimensional space. Therefore, we can say that the unit circle is a curve and the function $\gamma(t)$ is a *parametrization* of the curve.

A curve can have many different parametrizations. For example, the image of

$$\gamma(t) = \langle \cos(3t), \sin(3t) \rangle$$

is also the unit circle in the xy plane, so it is another parametrization of the same curve. The only difference between the two parametrizations is the speed at which they trace out the circle. The second function will go around the circle more rapidly than the first: at $t = \frac{2\pi}{3}$, the second parametrization has already traced out the full circle, while the first one hasn't even gone halfway.

Orientation

Often times we would like a *unit* vector which points tangent to a curve. We already know that the derivative of the vector function $\gamma'(t)$ will be a tangent vector, so to make it into a unit vector we just divide by the magnitude. We therefore define the **unit tangent vector** to a curve as

$$\mathbf{T}(t) = \frac{\gamma'(t)}{|\gamma'(t)|}$$

If two different parametrizations trace out the same curve *in the same direction* (we will talk more about this shortly), then their unit tangent vectors at a particular value of t will always be the same.

Example. Find the unit tangent vector at $t = 1$ for the following two parametrizations of the same parabola $y = x^2$

$$\gamma_1(t) = \langle t, t^2 \rangle \qquad \text{and} \qquad \gamma_2(t) = \langle t^3, t^6 \rangle$$

We start with the first one; all we need to do is find the derivative vector at $t = 1$ and turn it into a unit vector.

$$\gamma_1'(1) = \langle 1, 2t \rangle = \langle 1, 2 \rangle$$

$$|\gamma_1'(1)| = \sqrt{1 + 4} = \sqrt{5}$$

Therefore, the unit tangent vector will be

$$\mathbf{T}_1(1) = \langle \frac{1}{\sqrt{5}}, \frac{2}{\sqrt{5}} \rangle$$

We now repeat the process for the second parametrization

$$\gamma_2'(1) = \langle 3t^2, 6t^5 \rangle = \langle 3, 6 \rangle$$

$$|\gamma_2'(1)| = \sqrt{9 + 36} = 3\sqrt{5}$$

$$\mathbf{T}_2(1) = \langle \frac{1}{\sqrt{5}}, \frac{2}{\sqrt{5}} \rangle$$

So they are the same, as expected because both parametrizations trace the parabola in the same direction, i.e. as t increases they both move "to the right".

Now we want to look a little closer at what we mean by the "direction" in which a parametrization traces out a certain curve. For example, the function

$$\gamma(t) = \langle -t, t^2 \rangle$$

is yet another parametrization of the exact same parabola $y = x^2$. However, this parametrization is different in that as t increases, the parabola is being traced out from right to left, opposite of the above two parametrizations. Let's look at the unit tangent vector to this new parametrization at the same point $(1, 1)$.

On this parametrization, the point $(1, 1)$ corresponds to $t = -1$, so

$$\gamma'(-1) = \langle -1, -2 \rangle$$

$$\mathbf{T}(-1) = \langle -\frac{1}{\sqrt{5}}, -\frac{2}{\sqrt{5}} \rangle$$

The unit tangent vector for this "backwards" parametrization points in the exact opposite direction as that of the ones from before.

This motivates us to define the "direction" of a curve, which we will call the **orientation**, in terms of which way the unit tangent vector points. There are only two possible orientations for a curve, and the orientation is decided by the parametrization used to describe the curve. For example, we would say that the parametrizations

$$\gamma(t) = \langle t, t^2 \rangle \qquad \text{and} \qquad \gamma(t) = \langle -t, t^2 \rangle$$

induce *opposite* orientations on the parabola, since their unit tangent vectors always point in opposite directions.

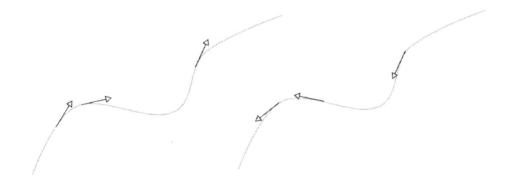

Figure 5.1: The same curve with two different parametrizations that induce opposite orientations on the curve. The arrows represent the unit tangent vectors at various points.

Arc Length

The last thing about curves that we need to look at is the concept of **arc length**. For example, say we have a curve in two dimensional space with parametrization $\gamma(t) = \langle x(t), y(t) \rangle$. We might want to know the length of the section of the curve traced out from $t = a$ to $t = b$.

We begin by taking a partition of t, i.e. splitting up the interval $[a, b]$ into many subintervals with length Δt. If we plot the points on the curve which correspond to the endpoints of all these subintervals, then we can connect these dots to form a bunch of straight lines which are pretty close to the actual curve. Adding up the lengths of all these straight lines will then give us an approximation to the actual arc length of this section of the curve.

Let's look at an arbitrary straight line which corresponds to the subinterval $[t_i, t_i + \Delta t]$. Again, the i's just remind us that we are looking at a random subinterval; these calculations apply to any of the subintervals. The straight line can be represented by the vector which goes between its endpoints

$$\gamma(t_i + \Delta t) - \gamma(t_i)$$

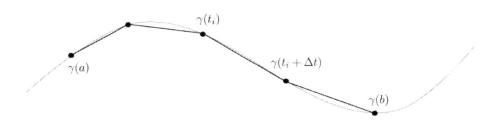

The magnitude of this vector will then be the approximate length of this little part of the curve. We can simplify the expression by writing out the components:

$$|\gamma(t_i + \Delta t) - \gamma(t_i)| = \sqrt{\left[x(t_i + \Delta t) - x(t_i)\right]^2 + \left[y(t_i + \Delta t) - y(t_i)\right]^2}$$

Now here comes the trick; the coordinate functions $x(t)$ and $y(t)$ are just single variable real valued functions, so we can invoke the *mean value theorem* from single variable calculus. This tells us that

$$x(t_i + \Delta t) - x(t_i) = x'(t_i^*)(\Delta t)$$

holds for some value of t_i^* in the interval $[t_i, t_i + \Delta t]$, and the same for $y(t)$. We can then rewrite the magnitude we are after as

$$\sqrt{\left[x(t_i + \Delta t) - x(t_i)\right]^2 + \left[y(t_i + \Delta t) - y(t_i)\right]^2} = \sqrt{\left[x'(t_i^*)\Delta t\right]^2 + \left[y'(t_i^*)\Delta t\right]^2}$$
$$= \sqrt{x'(t_i^*)^2 + y'(t_i^*)^2}\, \Delta t$$
$$= |\gamma'(t_i^*)|\, \Delta t$$

This tells us that the length of each straight line corresponding to some subinterval of t can be calculated by multiplying the size of the subinterval Δt by the magnitude of the derivative vector $\gamma'(t)$ evaluated at some point in the subinterval.

Adding up the lengths of all the straight lines will produce a *Riemann sum* which approximates the length of the curve, and as the number of subintervals increases the approximation will become more accurate. Therefore, we define the length of the curve from $t = a$ to $t = b$ by

$$\int_a^b |\gamma'(t)|\, dt$$

If you think of $\gamma(t)$ as the position vector at time t of a particle moving along the curve, then this formula simply says that the distance traveled is equal to the integral of speed $|\gamma'(t)|$ with respect to time.

Also note that even though we derived this equation for the arc length of a curve in two dimensional space, it works just as well for curves in three dimensional space.

Example. Find the length of the segment of the helix given by

$$\gamma(t) = \langle \cos t, \sin t, t \rangle \qquad 0 \le t \le 2\pi$$

All we have to do is find the magnitude of the derivative vector and integrate it.

$$\gamma'(t) = \langle -\sin t, \cos t, 1 \rangle$$

$$|\gamma'(t)| = \sqrt{\sin^2 t + \cos^2 t + 1} = \sqrt{2}$$

Therefore, the length of the arc is

$$\int_0^{2\pi} \sqrt{2}\, dt = 2\sqrt{2}\pi$$

Line Integrals

We have already seen double and triple integrals, where we integrate a function over a region in two or three dimensional space. Throughout the rest of the book, we will extend the concept of the integral to include even more types of domains and also different types of integrands. In this section, we look at integrals of real valued functions where the domain of integration is a *curve*.

We can define this new type of integral using Riemann sums, just like we did with double and triple integrals. The first step is to take a partition of our domain. Since our domain is a curve C in two or three dimensional space, we can take this to mean cutting up the curve into n little sections which each have an arc length of Δs.

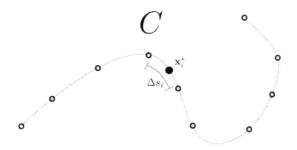

Next we take a random point within each little section $\mathbf{x}^*$ and find the value of the integrand function there $f(\mathbf{x}^*)$, then multiply this by the length of that little section. The last step is to add up this quantity for every little section of the curve, and let the number of little sections increase to infinity. We call this the ***line integral*** of the integrand function $f(\mathbf{x})$ over the curve C.

$$\int_C f(\mathbf{x})\, ds = \lim_{n \to \infty} \sum_{i=1}^n f(\mathbf{x}_i^*) \Delta s_i$$

You should notice that our construction of the Riemann sum follows the exact same process we used in earlier integrals. We split our domain into little pieces, multiply the size (arc length in this case) of each little piece by the value of the integrand at a random point in that piece, then add up the results across the entire domain.

Evaluating Line Integrals

Now we know what a line integral means, but we have no idea how to actually calculate one. It turns out that we can reduce a line integral to a regular old single integral, and the formula is quite intuitive. To start out, we need a parametrization $\gamma(t)$ of the curve we are integrating over.

In our Riemann sum for the line integral, the size of each little section of the curve was represented by its arc length Δs_i. Now suppose that this little section of the curve is traced out by our parametrization from $t = t_i$ to $t = t_i + \Delta t$. Then, using the formula from last section, the arc length of this part of the curve is given by

$$\Delta s_i = \int_{t_i}^{t_i+\Delta t} |\gamma'(t)|\ dt$$

By the mean value theorem for integrals, the following must be true for some value of t in the interval $[t_i, t_i + \Delta t]$

$$\Delta s_i = |\gamma'(t_i^*)|\,\Delta t$$

So you can think of all the Δs in our original Riemann sum getting replaced by Δt multiplied by the magnitude of the derivative vector of our parametrization.

Therefore, if the curve C we are integrating over is parametrized by $\gamma(t)$ from $t = a$ to $t = b$, then we can evaluate a line integral like so:

$$\int_C f(\mathbf{x})\,ds = \int_a^b f(\gamma(t))\,|\gamma'(t)|\ dt$$

So to transform our line integral to a single integral we must do the following: find a parametrization, multiply the integrand by the magnitude of the derivative vector of this parametrization, and lastly replace all the x, y, and z from our integrand in terms of t.

Note that the value of the line integral does *not* depend on the parametrization that we choose for the curve, even if two parametrizations induce opposite orientations. If you think about it, the arc length Δs of a little piece of the curve and the value of the integrand function at a point $f(\mathbf{x})$ should not depend on which parametrization we choose. We can describe the curve however we want, but it's still going to have the same length and our integrand will still have the same value at the same points. In a few sections, we will look at another case of line integrals where the orientation of the curve matters.

Example. Suppose we have a metal wire in the shape of a helix of radius 1. If the density in mass per length at a point (x, y, z) on the wire is given by $\delta(x, y, z) = x + y + z$, what is the mass of two turns of the wire?

Imagine dividing the wire into small sections, each with length Δs. If these sections are small enough, the density remains approximately constant on it, so the mass of each section is its length multiplied by the value of the density function at some point in the section. The total mass of the wire is then the sum of the masses of all the little sections and can be represented with the line integral

$$\int_C \delta(x, y, z)\, ds$$

To calculate this, we first need a parametrization of two turns of a helix. This is given by

$$\gamma(t) = \langle \cos t, \sin t, t \rangle \qquad 0 \le t \le 4\pi$$

Next, we need the magnitude of the derivative vector

$$\gamma'(t) = \langle -\sin t, \cos t, 1 \rangle$$

$$|\gamma(t)| = \sqrt{2}$$

All that's left is to rewrite the integrand in terms of t according to our parametrization, then evaluate the single integral

$$\int_C x + y + z\, ds = \int_0^{4\pi} \left(\cos t + \sin t + t \right) \sqrt{2}\, dt$$

$$= \sqrt{2} \left[\sin t - \cos t + \frac{1}{2} t^2 \right]_0^{4\pi}$$

$$= 8\sqrt{2}\pi^2$$

Exercises

1. Find a parametrization of the upper half of the circle of radius 2

    (a) Find the unit tangent vector at $(\sqrt{2}, \sqrt{2})$

    (b) Find another parametrization which induces an opposite orientation

    (c) Find the unit tangent vector at the same point and show that it is opposite to the one obtained in part (a)

2. Find the arc length of the given curve segment

    (a) $\gamma(t) = \langle t^2, \cos t + t \sin t, \sin t - t \cos t \rangle \quad 0 \le t \le 2\pi$

(b) $\gamma(t) = \langle t^2, t^3 \rangle \quad 0 \le t \le 1$

3. A fence is built along the curve $y = x^3$ from $(0,0)$ to $(2,8)$. The height at each point on the fence is given by $h(x, y) = y$. What is the surface area of the fence?

Answers

1. $\gamma(t) = \langle 2\cos t, 2\sin t \rangle$

   (a) $\mathbf{T}(\frac{\pi}{4}) = \langle -\frac{\sqrt{2}}{2}, \frac{\sqrt{2}}{2} \rangle$

   (b) $\gamma(t) = \langle -2\cos t, 2\sin t \rangle$

   (c) $\mathbf{T}(\frac{3\pi}{4}) = \langle \frac{\sqrt{2}}{2}, -\frac{\sqrt{2}}{2} \rangle$

2. (a) $2\sqrt{5}\pi^2$

   (b) $\frac{1}{27}\left[13\sqrt{13} - 8\right]$

3. Given by the line integral $\int_C h(x, y)ds$. Using the parametrization $\gamma(t) = \langle t, t^3 \rangle \quad 0 \le t \le 2$, we have

$$\int_C h(x, y)\, ds = \int_0^2 t^3 \sqrt{9t^4 + 1}\, dt = \frac{1}{54}(145\sqrt{145} - 1)$$

5.2   Surface Integrals

All the parametric stuff we've talked about so far has been about curves. The image of a single variable vector valued function $\gamma(t)$ is a curve, and we say that the function $\gamma(t)$ parametrizes that particular curve.

What happens if we look at the image of a *two* variable vector function? In particular, a function $r : \mathbb{R}^2 \mapsto \mathbb{R}^3$ with an output in three dimensional space. Just like we used t to denote a random parameter before, for two variable vector functions we will use u and v. So we are looking at the image of a function $r(u, v) = \langle x(t), y(t), z(t) \rangle$.

In general, the image of these kinds of functions will be a *surface*, where each point (x, y, z) on the surface corresponds to some values of the parameters u and v. To visualize this, imagine the uv plane being lifted up and thrown into three dimensional space in the shape of some surface. Compare this with one dimensional vector functions, where we imagined the t number line being picked up and twisted into a curve.

If a surface S is the image of some vector function $r(u, v)$, then we say that the function is a parametrization of the surface. As with parametric curves, we must keep in mind what our parameters u and v stand for in a particular function. Let's look at some examples.

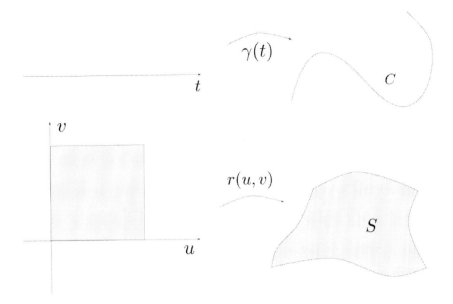

Example.
$$r(u,v) = \langle u, v, u^2 + v^2 \rangle$$

In this parametrization, u and v just stand for x and y, respectively. The z component always satisfies $z = x^2 + y^2$, so the surface parametrized by this function is just the paraboloid.

This is an example of a what happens when we parametrize the *graph* of a two variable function $z = f(x,y)$. We can just choose our two parameters to be x and y and then we get our z component for free since it is already expressed in terms of x and y through the function. As another example, to parametrize the graph of

$$f(x,y) = e^{xy} - 3y^2$$

we can just do

$$r(u,v) = \langle u, v, e^{uv} - 3v^2 \rangle$$

Example. Suppose we want to parametrize the unit sphere. We need to find two good parameters which we can use to describe any point on the sphere. The easiest way to do this is to think in spherical coordinates.

At all points on the surface, $\rho = 1$. We can therefore take our two parameters to be the θ and ϕ from spherical coordinates, i.e. the two angles that they represent. Any point on the sphere can be described by these two angles. For example, the point $(1,0,0)$ is described by $\theta = 0$ and $\phi = \pi/2$. Borrowing the conversion formulas between spherical and rectangular coordinates, we write our parametrization as

$$r(\theta,\phi) = \langle \sin\phi\cos\theta, \sin\phi\sin\theta, \cos\phi \rangle \quad 0 \le \theta \le 2\pi, \quad 0 \le \phi \le \pi$$

We could have used u and v to represent θ and ϕ, but if we can use notation that will help us remember what the parameters actually stand for that is usually better.

Surface Area

Now that we have these parametric surfaces, it is natural to ask how we can find their surface area. As with arc length for curves, surface area will be important in defining new types of integrals.

The derivation of the formula looks very similar to that of arc length and change of variables. The first thing we do is divide our domain in the uv plane into many little squares, just like we started out arc length by taking a partition of t.

Each of these little squares in the uv plane has sides of Δu and Δv and corresponds to a little patch of the parametric surface, so we want to find an expression for the area of one little patch. Just like in change of variables, we will assume that the patch corresponding to each square looks something like a parallelogram, so that we can compute its area with the magnitude of a cross product.

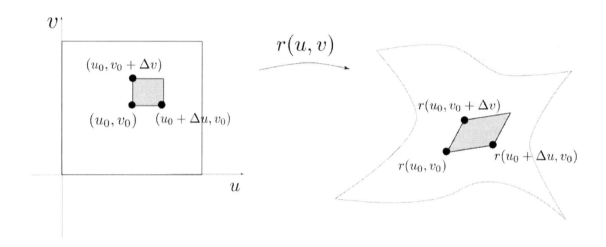

The two sides of each parallelogram can be represented by the vectors

$$r(u_0 + \Delta u, v_0) - r(u_0, v_0) \qquad \text{and} \qquad r(u_0, v_0 + \Delta v) - r(u_0, v_0)$$

If each square in the uv plane was small, then Δu and Δv are pretty small, so we can approximate these vectors by the differential of $r(u, v)$, which gets us

$$\frac{\partial r}{\partial u}\Delta u \qquad \text{and} \qquad \frac{\partial r}{\partial v}\Delta v$$

The area of each small parallelogram is then given by the magnitude of the cross product of these vectors

$$\left| \frac{\partial r}{\partial u} \times \frac{\partial r}{\partial v} \right| \Delta u \Delta v$$

To get the total surface area, we add up the areas of all the small parallelograms, which forms a double integral over our domain in the uv plane. Therefore, the parametric surface traced out as u and v vary through a domain D has surface area

$$\iint_D \left| \frac{\partial r}{\partial u} \times \frac{\partial r}{\partial v} \right| dA$$

Example. Find the lateral surface area (the area of the side) of a cylinder of radius a and height h. You obviously already know the answer is $2\pi ah$, but we can get that through an integral to illustrate how the formula is used.

First we'll need a parametrization of the cylinder. The easiest one comes by thinking in cylindrical coordinates. The r from cylindrical coordinates has a constant value of a at all points on the cylinder. We can then choose our two parameters to be θ and z and use the conversion formulas between cylindrical and rectangular coordinates:

$$r(\theta, z) = \langle a\cos\theta, a\sin\theta, z \rangle$$

To calculate the surface area, we need the magnitude of the cross product of the partial derivatives of this parametrization

$$\frac{\partial r}{\partial \theta} = \langle -a\sin\theta, a\cos\theta, 0 \rangle$$

$$\frac{\partial r}{\partial z} = \langle 0, 0, 1 \rangle$$

$$\frac{\partial r}{\partial \theta} \times \frac{\partial r}{\partial z} = \langle a\cos\theta, a\sin\theta, 0 \rangle$$

$$\left| \frac{\partial r}{\partial \theta} \times \frac{\partial r}{\partial z} \right| = a$$

Therefore, we calculate the surface area with the following integral, noting that θ ranges from 0 to 2π and z ranges from 0 to the height h of the cylinder

$$\iint_D a\, dA = \int_0^{2\pi} \int_0^h a\, dA$$

$$= 2\pi ah$$

Orientation

Recall that we defined orientation for curves by the way the unit tangent vector pointed. If two parametrizations of the same curve had opposite orientations, then that meant at every point their unit tangent vectors point in opposite directions.

We would like a similar way to define the orientation that a particular parametrization induces on a surface. Since tangent vectors don't seem very useful for this purpose, we look to vectors which are normal to the surface.

Let's go back to the cross product we used to derive the surface area formula:

$$\frac{\partial r}{\partial u} \times \frac{\partial r}{\partial v}$$

Both partial derivative vectors lie inside the surface, so their cross product must be perpendicular to the surface. We therefore define the **unit normal vector** to a parametric surface as

$$\mathbf{N}(u, v) = \frac{\dfrac{\partial r}{\partial u} \times \dfrac{\partial r}{\partial v}}{\left| \dfrac{\partial r}{\partial u} \times \dfrac{\partial r}{\partial v} \right|}$$

At any given point on a surface, there are only two possible unit normal vectors, and they point in opposite directions. If two different parametrizations of the same surface induce opposite orientations, then at every point their unit normal vectors will be opposite.

Surface Integrals

We can now define a new type of integral; this time we will be integrating a real valued function over a surface. The process of defining and finding a way to compute these closely resembles what we did last section for line integrals, so we will go over this quickly.

First, we split our surface into many small patches, each with a surface area ΔS. For each small patch, we choose a random point in the patch (x^*, y^*, z^*) and evaluate the integrand function at that point $f(x^*, y^*, z^*)$. Multiply this by the surface area of the patch, then sum up this quantity over all the patches which make up the surface, and finally let the number of little patches increase. We call this a ***surface integral***.

$$\iint_S f(x, y, z)\, dS = \lim_{n \to \infty} \sum_{i=1}^{n} f(x_i^*, y_i^*, z_i^*) \Delta S_i$$

To actually compute a surface integral, we first need a parametrization of the surface. Just like with line integrals, whatever parametrization of the surface we choose has no effect on the surface integral. When deriving the formula for surface area, we found that a small square in the uv plane with area

$$\Delta A = \Delta u \Delta v$$

corresponds to a parallelogram like shape on the parametric surface with surface area

$$\left| \frac{\partial r}{\partial u} \times \frac{\partial r}{\partial v} \right| \Delta u \Delta v$$

Intuitively, you can think that all the little ΔS in our Riemann sum for a surface integral are being replaced by this quantity. It then makes sense to use

$$\iint_S f(x, y, z)\, dS = \iint_D f(r(u, v)) \left| \frac{\partial r}{\partial u} \times \frac{\partial r}{\partial v} \right| dA \qquad .$$

as a way to compute surface integrals, where $r(u, v)$ is our parametrization of the surface and D is the two dimensional domain that the parameters u and v range over.

Example. Compute the surface integral of $f(x, y, z) = xy$ over the part of the paraboloid $z = x^2 + y^2$ which lies over the square in the xy plane $0 \leq x \leq 1, 0 \leq y \leq 1$.

First, we find a parametrization of the surface; remember that for any graph of a function $z = f(x, y)$, choosing x and y as the parameters is always an easy choice.

$$r(u, v) = \langle u, v, u^2 + v^2 \rangle$$

Next, we find the magnitude of the cross product that we need for the surface integral

$$\frac{\partial r}{\partial u} = \langle 1, 0, 2u \rangle$$

$$\frac{\partial r}{\partial v} = \langle 0, 1, 2v \rangle$$

$$\frac{\partial r}{\partial u} \times \frac{\partial r}{\partial v} = \langle -2u, -2v, 1 \rangle$$

$$\left| \frac{\partial r}{\partial u} \times \frac{\partial r}{\partial v} \right| = \sqrt{4u^2 + 4v^2 + 1}$$

Since we chose $u = x$ and $v = y$, our domain is just going to correspond to the same square in the uv plane. Therefore, the surface integral is

$$\iint_S xy \, dS = \iint_D uv \sqrt{4u^2 + 4v^2 + 1} \, dA$$

$$= \int_0^1 \int_0^1 uv \sqrt{4u^2 + 4v^2 + 1} \, du \, dv$$

Exercises

1. Find a parametrization of the surface

   (a) The plane $x - y + z = 2$

   (b) The part of the cone $z = \sqrt{x^2 + y^2}$ below the plane $z = 4$

2. Use an integral to find the surface area of a sphere of radius a

3. A metal sheet is in the shape of the plane $3x + 2y + z = 1$ over the square $0 \leq x \leq 1$, $0 \leq y \leq 1$ in the xy plane. The density at a point on the sheet in mass per surface area is given by $\delta(x, y, z) = xy$. What is the total mass?

Answers

1. (a) Choosing x and y as parameters, $r(u, v) = \langle u, v, 2 + v - u \rangle$

   (b) Choosing r and θ from cylindrical coordinates as parameters,

   $$r(r, \theta) = \langle r \cos \theta, r \sin \theta, r \rangle \quad 0 \leq r \leq 4, 0 \leq \theta \leq 2\pi$$

2. Use parametrization $r(\theta, \phi) = \langle a \sin \phi \cos \theta, a \sin \phi \sin \theta, a \cos \phi \rangle$, the integral is

$$\int_D a^2 \sin \phi \, dA = \int_0^{2\pi} \int_0^{\pi} a^2 \sin \phi \, d\phi \, d\theta = 4\pi a^2$$

3. $\frac{\sqrt{14}}{4}$; Use parametrization $r(u, v) = \langle u, v, 1 - 3u - 3v \rangle$, then

$$\iint_S \delta(x, y, z) \, dS = \iint_D \sqrt{14} uv \, dA = \sqrt{14} \int_0^1 \int_0^1 uv \, du \, dv$$

5.3 Vector Fields

For the rest of the book, we will be integrating a whole different type of function, as opposed to the real valued functions which have been our integrands so far. A **vector field** is just a function which assigns a vector to every point. The dimensions of the input and output of a vector field are always the same. For example, say we have the vector field

$$\mathbf{F}(x, y) = \langle x, y \rangle$$

We plug a point into the function, and it gives us a vector to put at that point. So since

$$\mathbf{F}(1, 2) = \langle 1, 2 \rangle$$

we place a vector $\langle 1, 2 \rangle$ with its tail at the point $(1, 2)$ on the xy plane. The full picture of this particular vector field looks like this:

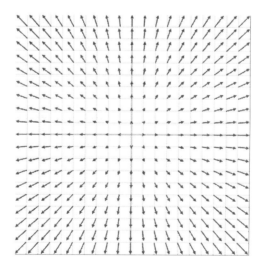

Vector fields usually have some sort of physical meaning associated with them. Perhaps we are studying wind patterns in an area, and the vector at each point represents the velocity vector of the wind there. Maybe we are looking at the Earth's gravitational field, and the arrow at each point stands for the force vector that an object would feel if it were placed there.

Another example of a vector field is the gradient of a real valued function. Say we have the function $f(x, y, z) = x^2y + z^3$, then its gradient is

$$\nabla f(x, y, z) = \langle 2xy, x^2, 3z^2 \rangle$$

When we plug in a point, we get a vector of the same dimensions back. In this case, the output vector represents the direction of fastest increase of the function at that point.

The coordinate functions of a vector field are commonly denoted by P, Q, and R. So for a three dimensional vector field,

$$\mathbf{F}(x, y, z) = \langle P(x, y, z), Q(x, y, z), R(x, y, z) \rangle$$

and for a two dimensional vector field,

$$\mathbf{F}(x, y) = \langle P(x, y), Q(x, y) \rangle$$

For the rest of this section, we will go over some terminology and computations associated with vector fields that will be used later on.

Conservative Fields

A vector field is said to be ***conservative*** if it is the gradient vector of some real valued function, which we call the *potential function*. For example,

$$\mathbf{F}(x, y) = \langle x, y \rangle$$

is a conservative vector field because we can find a potential function for it, such as

$$\varphi(x, y) = \frac{1}{2}x^2 + \frac{1}{2}y^2$$

$$\nabla \varphi(x, y) = \langle x, y \rangle$$

Example. Determine if the vector field is conservative

$$\mathbf{F}(x, y) = \langle 3x^2y + 3y, x^3 + 3x + 2y \rangle$$

If it is conservative, then there exists a real valued function $\varphi(x, y)$ whose gradient is equal to it, which means

$$\frac{\partial \varphi}{\partial x} = 3x^2y + 3y \qquad \text{and} \qquad \frac{\partial \varphi}{\partial y} = x^3 + 3x + 2y$$

Let's focus on the first component of the vector field; this is what you obtain by partial differentiating the potential function with respect to x. Therefore, if we integrate the first component with respect to x while treating y as a constant, we should be able to recover the original function.

$$\varphi(x,y) = \int 3x^2y + 3y\, dx = x^3y + 3xy + g(y)$$

Notice that instead of a plain constant C, we now have an arbitrary function of y. Why? Remember that we add C when taking an indefinite integral because any constants would have disappeared when we differentiated. In this case, during partial differentiation with respect to x, any term only involving y would have been lost as well, so we must add this unknown function of y.

We now need to determine what $g(y)$ is. To do this, we utilize the fact that differentiating what we have now with respect to y must be equal to the second component of the vector field.

$$\frac{\partial\varphi}{\partial y} = x^3 + 3x + g'(y) = x^3 + 3x + 2y$$

Therefore,

$$g'(y) = 2y$$

$$g(y) = y^2 + C$$

So, in the end, this vector field is conservative and its potential function is given by

$$\varphi(x,y) = x^3y + 3xy + y^2 + C$$

Note that we could have gotten the same answer by focusing on the second component first. We would have integrated the second component with respect to y, making sure to add on an unknown function of x at the end. Then we would differentiate that with respect to x and equate it to the first component of the vector field to find out that unknown function.

We don't want to waste time trying to find a potential function if it isn't going to exist, and fortunately there is a quick way to check whether or not a vector field is conservative. A two dimensional vector field is conservative if

$$\frac{\partial P}{\partial y} = \frac{\partial Q}{\partial x}$$

The big idea here is the equality of mixed second partial derivatives. If the vector field is conservative with potential function $\varphi(x,y)$, it can be written as

$$\mathbf{F}(x,y) = \langle \frac{\partial\varphi}{\partial x}, \frac{\partial\varphi}{\partial y} \rangle$$

so therefore

$$\frac{\partial P}{\partial y} = \frac{\partial^2\varphi}{\partial y\partial x} = \frac{\partial^2\varphi}{\partial x\partial y} = \frac{\partial Q}{\partial x}$$

Similar conditions for conservative three dimensional vector fields can be found, again utilizing the equality of mixed partials.

Curl and Divergence

There are two operations on vector fields which will come up again. The **_divergence_** of a vector field is defined as

$$\text{div } \mathbf{F} = \frac{\partial P}{\partial x} + \frac{\partial Q}{\partial y} + \frac{\partial R}{\partial z}$$

For a two dimensional vector field, the definition is the same except the third term will be missing. So the divergence takes a vector field and turns it into a scalar real valued function. The divergence of a vector field at a point measures how much the vector field is flowing outwards from that point. For example, consider the two vector fields

$$\mathbf{F}_1(x, y) = \langle x, y \rangle \qquad \text{and} \qquad \mathbf{F}_2(x, y) = \langle -x, -y \rangle$$

$$\text{div } \mathbf{F}_1 = 2 \qquad \text{and} \qquad \text{div } \mathbf{F}_2 = -2$$

In the first vector field, shown on the first page of this section, the vectors are very clearly going outward from every point. In the second vector field, the vectors are all converging inward to the origin, so the divergence is negative.

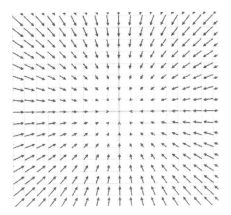

A way to remember the divergence is by the expression

$$\text{div } \mathbf{F} = \nabla \cdot \mathbf{F}$$

where ∇ is supposed to be the vector

$$\nabla = \langle \frac{\partial}{\partial x}, \frac{\partial}{\partial y}, \frac{\partial}{\partial z} \rangle$$

Again, note that this is just a way to remember the divergence, as you can't just multiply a partial derivative sign by a function to differentiate it.

The second operation is the **_curl_** of a vector field. The curl takes a three dimensional vector field and produces a new vector field.

$$\text{curl } \mathbf{F} = \left\langle \frac{\partial R}{\partial y} - \frac{\partial Q}{\partial z}, \frac{\partial P}{\partial z} - \frac{\partial R}{\partial x}, \frac{\partial Q}{\partial x} - \frac{\partial P}{\partial y} \right\rangle$$

You can remember this by the expression

$$\text{curl } \mathbf{F} = \nabla \times \mathbf{F}$$

Note that even though curl is defined for a three dimensional vector field, we can take the curl of a two dimensional one by putting $R = 0$, just like how we can take the cross product of two dimensional vectors.

The curl of a vector field at a point measures how much the vectors are circulating around that point. The curl vector will point in the direction of the axis of rotation and its magnitude will indicate how much swirling action is going on. For example, consider the vector field

$$\mathbf{F} = \langle -y, x \rangle$$

$$\text{curl } \mathbf{F} = \langle 0, 0, 2 \rangle$$

The curl therefore points along the z axis, indicating that the vectors are swirling around in the xy plane.

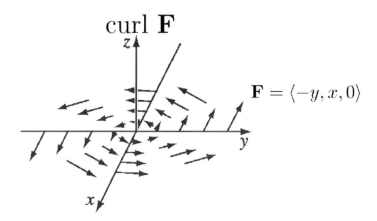

Also notice that the divergence of this vector field is 0, which makes sense because the vectors are not going outward or inward; they just circle around.

Exercises

1. Consider the vector field $\mathbf{F}(x, y) = \langle 9x^2 + 2y, 2x + 2y \rangle$

   (a) Test if the vector field is conservative

   (b) Find the potential function

2. Develop a test to determine whether or not a three dimensional vector field is conservative, based on the equality of mixed partial derivatives

   (a) Test whether the vector field $\mathbf{F}(x, y, z) = \langle 6xy, 3x^2 + 2z^2, 4yz \rangle$ is conservative

(b) Find the potential function

3. Find the divergence and curl of the vector field

$$\mathbf{F}(x, y, z) = \langle xyz, y^3 + 2z, 3x^2 + 4y \rangle$$

4. Prove that the curl of a conservative vector field is the zero vector

5. Prove that the divergence of the curl of a vector field is zero

Answers

1. (a) $\frac{\partial P}{\partial y} = 2 = \frac{\partial Q}{\partial x}$

   (b) $\varphi(x, y) = 3x^3 + 2xy + y^2 + C$

2.
$$\frac{\partial P}{\partial y} = \frac{\partial Q}{\partial x}$$
$$\frac{\partial P}{\partial z} = \frac{\partial R}{\partial x}$$
$$\frac{\partial Q}{\partial z} = \frac{\partial R}{\partial y}$$

   (a) Use the conditions from the previous part of the question

   (b) $\varphi(x, y, z) = 3x^2y + 2yz^2$

3.
$$\text{div } \mathbf{F} = yz + 3y^2$$
$$\text{curl } \mathbf{F} = \langle 2, xy - 6x, -xz \rangle$$

4. Use the equality of mixed partials, e.g.
$$\mathbf{F} = \langle \frac{\partial \varphi}{\partial x}, \frac{\partial \varphi}{\partial y}, \frac{\partial \varphi}{\partial z} \rangle$$
$$\text{curl } \mathbf{F} = \langle \frac{\partial^2 \varphi}{\partial y \partial z} - \frac{\partial^2 \varphi}{\partial z \partial y}, ... \rangle$$

5. Again, it involves the equality of mixed partials
$$\text{curl } \mathbf{F} = \left\langle \frac{\partial R}{\partial y} - \frac{\partial Q}{\partial z}, \frac{\partial P}{\partial z} - \frac{\partial R}{\partial x}, \frac{\partial Q}{\partial x} - \frac{\partial P}{\partial y} \right\rangle$$
$$\text{div curl } \mathbf{F} = \frac{\partial^2 R}{\partial x \partial y} - \frac{\partial^2 Q}{\partial x \partial z} + \frac{\partial^2 P}{\partial y \partial z} - \frac{\partial^2 R}{\partial y \partial x} + \frac{\partial^2 Q}{\partial z \partial x} - \frac{\partial^2 P}{\partial z \partial y} = 0$$

5.4   Integrals Over Vector Fields

Most applications of line and surface integrals involve integrating a *vector field*. When we developed these integrals, however, we only used real valued functions as integrands. In this section we will define what it means to take the line or surface integral of a vector field.

Line Integrals

Remember how we defined line integrals: split our curve into little sections, multiply the length of each section Δs by the value of the integrand function at a point within that section, and add them all up. We can use this same approach if we can somehow turn a vector field into a real valued function. The answer lies in the *dot product*.

Many times we are only interested in the component of a vector field which is *parallel* to the curve we are integrating over. The most common example of this is work, where we only want the component of force which is in the direction of the displacement. This suggests that, instead of trying to integrate a vector field itself, we should integrate

$$\mathbf{F} \cdot \mathbf{T}$$

the dot product of the vector field and the unit tangent vector. So in each little section of the curve, we multiply its length Δs by the value of the vector field dot the unit tangent vector at a point within the section. This is what we mean by the line integral of a vector field

$$\int_C \mathbf{F} \cdot \mathbf{T} \, ds$$

There is a convenient way to calculate these without actually finding the unit tangent vector. Remember to evaluate line integrals we need a parametrization of the curve C, and we multiply the integrand by the magnitude of the derivative of this parametrization.

$$\int_C \mathbf{F} \cdot \mathbf{T} \, ds = \int_a^b \mathbf{F}(\gamma(t)) \cdot \frac{\gamma'(t)}{|\gamma'(t)|} \, |\gamma'(t)| \, dt$$
$$= \int_a^b \mathbf{F}(\gamma(t)) \cdot \gamma'(t) \, dt$$

Therefore, to evaluate a line integral of a vector field we just dot it with the derivative of the parametrization, making sure to express everything in terms of t.

Also note that, unlike line integrals of regular scalar functions, line integrals of vector fields depend on the orientation of the curve. If we evaluate the same line integral two times using parametrizations which induce opposite orientations on the curve, then our answers will be opposites of each other. If we let C stand for the curve and $-C$ stand for the same curve with an opposite orientation, then

$$\int_C \mathbf{F} \cdot \mathbf{T} \, ds = - \int_{-C} \mathbf{F} \cdot \mathbf{T} \, ds$$

Intuitively this makes sense because an opposite orientation of the curve makes the unit tangent vector point in the opposite direction at every point along the curve, i.e. it gets multiplied by -1.

Example. Evaluate the line integral of the vector field $\mathbf{F}(x, y) = \langle x, y \rangle$ over the part of the parabola $y = x^2$ between $(0,0)$ and $(2,4)$.

First, we need to parametrize the curve.

$$\gamma(t) = \langle t, t^2 \rangle \qquad 0 \le t \le 2$$

Then

$$\begin{aligned}
\int_C \mathbf{F} \cdot \mathbf{T} \, ds &= \int_0^2 \mathbf{F}(\gamma(t)) \cdot \gamma'(t) \, dt \\
&= \int_0^2 \langle t, t^2 \rangle \cdot \langle 1, 2t \rangle \, dt \\
&= \int_0^2 t + 2t^3 \, dt \\
&= \left[\frac{1}{2} t^2 + \frac{1}{2} t^4 \right]_0^2 = 10
\end{aligned}$$

Let's see what happens if we integrate along the curve in the other direction, that is, using a different parametrization which induces an opposite orientation.

$$\gamma(t) = \langle -t, t^2 \rangle \qquad -2 \le t \le 0$$

The line integral will be opposite, as expected:

$$\begin{aligned}
\int_{-C} \mathbf{F} \cdot \mathbf{T} \, ds &= \int_{-2}^0 \langle -t, t^2 \rangle \cdot \langle -1, 2t \rangle \, dt \\
&= \int_{-2}^0 t + 2t^3 \, dt \\
&= -10
\end{aligned}$$

Surface Integrals

Next we can define what we mean by the integral of a vector field over a surface. For line integrals, we took the dot product of the vector field with the unit tangent vector, which determines orientation for curves. Since the unit *normal* vector determines orientation for a surface, it seems natural to define a surface integral of a vector field as

$$\iint_S \mathbf{F} \cdot \mathbf{N} \, dS$$

So we are only interested in the component of each vector which is perpendicular to the surface. These kinds of surface integrals are often used to compute how much of something is flowing out of the surface. For example, if the vector field represents the velocity vectors of some water, then the surface integral computes how much water is going out of the surface. For this reason, these kinds of surface integrals of vector fields are often called *flux* integrals.

Again, there is an easy way to compute these if we have a parametrization $r(u,v)$ of the surface.

$$\iint_S \mathbf{F} \cdot \mathbf{N}\, dS = \iint_D \mathbf{F}(r(u,v)) \cdot \frac{\dfrac{\partial r}{\partial u} \times \dfrac{\partial r}{\partial v}}{\left| \dfrac{\partial r}{\partial u} \times \dfrac{\partial r}{\partial v} \right|} \left| \frac{\partial r}{\partial u} \times \frac{\partial r}{\partial v} \right| dA$$

$$= \iint_D \mathbf{F}(r(u,v)) \cdot \left(\frac{\partial r}{\partial u} \times \frac{\partial r}{\partial v} \right) dA$$

Also, just like with line integrals of vector fields, reversing the orientation of the surface will reverse the sign of the surface integral.

Example. Calculate the integral of $\mathbf{F}(x,y,z) = \langle x,y,z \rangle$ over the part of the paraboloid $z = x^2 + y^2$ which lies over the circle of radius 2 in the xy plane and is *oriented downward*.

An easy parametrization of the surface is given by

$$r(u,v) = \langle u, v, u^2 + v^2 \rangle$$

where u and v range over the circle $u^2 + v^2 = 4$ in the uv plane, since they correspond to x and y.

$$\iint_S \mathbf{F} \cdot \mathbf{N}\, dS = \iint_D \mathbf{F}(r(u,v)) \cdot \left(\frac{\partial r}{\partial u} \times \frac{\partial r}{\partial v} \right) dA$$

$$= \iint_D \langle u, v, u^2 + v^2 \rangle \cdot \langle -2u, -2v, 1 \rangle\, dA$$

$$= \iint_D -(u^2 + v^2)\, dA$$

Now we must pay attention to the words "oriented downward" in the question, which tell us that the unit normal vector at every point on the surface should be pointing down. The unit normal vector given by our parametrization

$$\mathbf{N}(u,v) = \langle -2u, -2v, 1 \rangle$$

points up because of its positive z component. So the parametrization we chose gave the opposite orientation of the surface that we want. Instead of trying to find a new parametrization, we can simply remember to multiply the value of our integral by -1.

$$\iint_D u^2 + v^2\, dA$$

We can then transform our integral to polar coordinates for evaluation.

$$\int_0^{2\pi} \int_0^2 r^3 \, dr \, d\theta = 8\pi$$

Exercises

1. Compute the line integral

    (a) $\mathbf{F}(x, y) = \langle -y, x \rangle$ over the top half of the unit circle, going counterclockwise

    (b) $\mathbf{F}(x, y) = \langle 4xy, 2x^2 + 3y^2 \rangle$ over the square $0 \leq x \leq 1$, $0 \leq y \leq 1$ going counterclockwise

2. Consider the surface $4x - 3y + z = 5$

    (a) Parametrize the surface

    (b) Find a normal vector (not unit normal) at each point

    (c) Compute the surface integral of $\mathbf{F}(x, y, z) = \langle xy, 2z, x^2 \rangle$ over the part of this plane which lies over the square $0 \leq x \leq 1$, $0 \leq y \leq 1$ with *downward* orientation

Answers

1. (a) π

    (b) 0; Split up into four line integrals, one for each edge

2. (a) $r(u, v) = \langle u, v, 5 + 3v - 4u \rangle$

    (b) $\langle 4, -3, 1 \rangle$

    (c) $\frac{77}{3}$; the normal vector from part (b) is up, so we work through the integral and multiply by -1 at the end

5.5   The Gradient Theorem

The Fundamental Theorem of Calculus

In the next two sections, we will be looking at four important theorems which can all be considered generalizations of the fundamental theorem of calculus to higher dimensions. What do we mean by that? You know by now (hopefully) that the fundamental theorem of calculus says

$$\int_a^b f'(x) \, dx = f(b) - f(a)$$

we can compute the single integral of a derivative of some function by evaluating the original function at the end points.

The domain of a single integral is an interval on the number line $[a, b]$ which we will call I. The "boundary" of a one dimensional domain can be thought of as just the endpoints a and b. To denote the boundary of a domain, we use the ∂ symbol; so the endpoints of this interval I can be written as ∂I.

If we use this new notation and think of an integral over two points as simply evaluating the function at those points, then the fundamental theorem of calculus becomes

$$\int_I f'(x)\, dx = \int_{\partial I} f(x)\, dx$$

The big idea is that on the left we have an integral of a derivative over some domain, an interval in this case, and on the right we have an integral of the original function over the *boundary* of that domain.

In fact, all the theorems in the next two sections, as well as the fundamental theorem of calculus, are just special cases of a very general statement called ***Stoke's Theorem***, which says

$$\int_D d\omega = \int_{\partial D} \omega$$

The actual statement uses something called differential forms and is too complicated to talk about in this book, but the big picture remains the same.

On the left, some kind of derivative (not the usual derivative we are used to) is being integrated on a domain. On the right, the original function is being integrated over the boundary of this domain. This relationship is useful because often it is easier to perform one integral than the other. As we examine the various theorems of vector calculus, see if you can spot this theme in the formulas.

The Gradient Theorem

The first theorem we look at is the direct analog of the fundamental theorem of calculus to a curve in two or three dimensional space. In fact, it is often called ***The Fundamental Theorem of Line Integrals***.

In section 5.3, we saw that the gradient of a real valued function is a vector field, and that if a vector field can be expressed as the gradient of some potential function, then it is called conservative.

The Gradient Theorem says that if we integrate a conservative vector field over a curve, then it can be calculated by evaluating the potential function at the endpoints of the curve.

$$\int_C \nabla f \cdot \mathbf{T}\, ds = f(\gamma(b)) - f(\gamma(a))$$

where $\gamma(t)$ is a parametrization of the curve C, and $\gamma(a)$ and $\gamma(b)$ correspond to the endpoints of the curve.

Proof.

$$\int_C \nabla f \cdot \mathbf{T}\, ds = \int_a^b \nabla f \cdot \gamma'(t)\, dt$$

$$= \int_a^b \frac{d}{dt}\left[f(\gamma(t))\right] dt \quad \text{(chain rule)}$$

$$= \left[f(\gamma(t))\right]_a^b \quad \text{(fundamental theorem of calculus)}$$

$$= f(\gamma(b)) - f(\gamma(a))$$

$\square$

Example. Calculate the line integral of $\mathbf{F}(x,y) = \langle 4y^3 + 2xy, 12xy^2 + x^2 \rangle$ over the parabola $y = x^2$ from $(1,1)$ to $(2,4)$. We could do it the old fashioned way, or we could recognize that this vector field is conservative because

$$\frac{\partial P}{\partial y} = 12y^2 + 2x = \frac{\partial Q}{\partial x}$$

and apply the Gradient Theorem. To do this we first need to find the potential function; we can integrate the first component with respect to x

$$\varphi(x,y) = \int 4y^3 + 2xy\, dx = 4xy^3 + x^2 y + g(y)$$

remembering to add our arbitrary function of y. Differentiating this with respect to y and equating it with the second component of the vector field reveals that $g(y)$ is zero. Therefore,

$$\int_C \mathbf{F} \cdot \mathbf{T}\, ds = \int_C \nabla\varphi(x,y) \cdot \mathbf{T}\, ds$$

$$= \varphi(2,4) - \varphi(1,1)$$

$$= 528 - 5 = 523$$

Notice that we didn't even need to find a parametrization of the curve because the endpoints would be the same no matter what parametrization we chose, provided they had the same orientation.

One important consequence of the Gradient Theorem is that the line integral of a conservative vector field does not depend on the curve you integrate over as long as they all have the same endpoints. So using the previous example, any curve that started at $(1,1)$ and ended at $(2,4)$, not just the parabola we used, would have given the same value of the line integral.

Also consider what happens if we try to integrate over a closed curve, meaning the curve connects with itself to form a loop, e.g. a circle. There are no clear endpoints of such a curve, so we could choose any point to be both the starting and ending point.

Therefore, the Gradient Theorem says that the line integral of a conservative vector field over a closed curve must be zero, since the values of the potential function at both endpoints are obviously the same if the endpoints are the same. In symbols,

$$\oint_C \nabla f \cdot \mathbf{T}\, ds = 0$$

where the integral with a circle in it indicates that the curve is closed.

Exercises

1. Consider the vector field $\mathbf{F}(x, y) = \langle 2y^2 + 6x, 4xy + 1 \rangle$

    (a) Show that the vector field is conservative, and find a potential function

    (b) Calculate the line integral over the parabola $y = x^2$ from $0 \le x \le 2$

    (c) Calculate the line integral over the line $y = 2x$ from the origin to the point where $y = 4$

    (d) Calculate the line integral over the unit circle

2. Calculate the line integral of $\mathbf{F}(x, y, z) = \langle 4x, 1, 3 \rangle$ over the curve

$$\gamma(t) = \langle t, t^2, t^3 \rangle \quad 0 \le t \le 2$$

    using both the normal method and the Gradient Theorem, then check that your answers are the same

Answers

1. (a) $\varphi(x, y) = 2xy^2 + 3x^2 + y + C$

    (b) 80

    (c) 80

    (d) 0

2. 36

5.6 The Theorems of Vector Calculus

Green's Theorem

Green's Theorem relates a double integral over a two dimensional region D to the line integral over its boundary curve ∂D. Before stating the theorem, we must define what we mean by a "positively oriented" boundary curve. If the boundary curve to a region is positively oriented, it means that if you walk along the curve in the direction its oriented in, the region should always be on your left. The following shows a region with a hole in it and the corresponding positively oriented boundary curves.

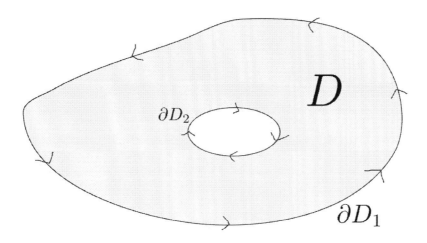

Green's Theorem says that if D is a two dimensional region and ∂D is its positively oriented boundary curve, then

$$\iint_D \left(\frac{\partial Q}{\partial x} - \frac{\partial P}{\partial y} \right) dA = \oint_{\partial D} \mathbf{F} \cdot \mathbf{T} \, ds$$

Example. Calculate the line integral of $\mathbf{F}(x, y) = \langle x^2 y, 3x \rangle$ over the triangle shown.

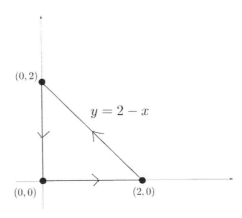

To actually compute the line integral, we would need to split it into three separate line integrals: one for each side of the triangle. However, Green's Theorem lets us instead calculate a double integral over the region enclosed by the triangle.

$$\oint_{\partial D} \mathbf{F} \cdot \mathbf{T} \, ds = \iint_D \left(\frac{\partial Q}{\partial x} - \frac{\partial P}{\partial y} \right) dA$$

$$= \iint_D 3 - x^2 \, dA$$

$$= \int_0^2 \int_0^{2-x} 3 - x^2 \, dy \, dx$$

$$= \int_0^2 x^3 - 2x^2 - 3x + 6 \, dx$$

$$= \frac{14}{3}$$

Sometimes the line integral is easier to compute than the double integral. Recall that the double integral of 1 just gives you the area of the domain of integration. Therefore, if we integrate a vector field where

$$\frac{\partial Q}{\partial x} - \frac{\partial P}{\partial y} = 1$$

over a closed curve, then it will give the area of the region enclosed by the curve. Examples of such vector fields include $\mathbf{F} = \langle 0, x \rangle$, $\mathbf{F} = \langle -y, 0 \rangle$, and $\mathbf{F} = \langle \frac{1}{2} x, -\frac{1}{2} y \rangle$.

Example. Use a line integral to find the area enclosed by the ellipse

$$\frac{x^2}{a^2} + \frac{y^2}{b^2} = 1$$

Earlier we saw that we could evaluate the integral $\iint_D dA$ over this domain with a change of variables, but it is much simpler to use Green's Theorem. The boundary of the ellipse region can be parametrized by

$$\gamma(t) = \langle a \cos t, b \sin t \rangle \qquad 0 \le t \le 2\pi$$

Therefore, Green's Theorem gives

$$\iint_D dA = \oint_{\partial D} \mathbf{F} \cdot \mathbf{T} \, ds$$

$$= \int_0^{2\pi} \langle 0, x \rangle \cdot \langle -a \sin t, b \cos t \rangle \, dt$$

$$= \int_0^{2\pi} ab \cos^2 t \, dt$$

$$= ab \left[\frac{1}{2} t + \frac{1}{4} \sin 2t \right]_0^{2\pi} = \pi ab$$

The Curl Theorem

The Curl Theorem is essentially the three dimensional case of Green's Theorem; it relates the surface integral of the curl of a vector field to the line integral of the vector field along the surface's boundary curves.

$$\iint_S \operatorname{curl} \mathbf{F} \cdot \mathbf{N} \, dS = \oint_{\partial S} \mathbf{F} \cdot \mathbf{T} \, ds$$

Example. Compute the line integral of $\mathbf{F}(x, y, z) = \langle x, x+y, x+y+z \rangle$ over the curve of intersection between the plane $z = y$ and the cylinder $x^2 + y^2 = 1$.

This slanted circle can be viewed as the boundary curve to the part of the plane $z = y$ which lies over the unit circle in the xy plane. Therefore, the Curl Theorem says that we can compute the surface integral of the curl of $\mathbf{F}$ over this surface rather than do the actual line integral.

$$\operatorname{curl} \mathbf{F} = \langle 1, -1, 1 \rangle$$

To find a normal vector to the surface, we could parametrize the plane and then compute

$$\frac{\partial r}{\partial u} \times \frac{\partial r}{\partial v}$$

but we could also get a normal vector directly from the equation of the plane. Recall that the coefficients of x, y, and z determine a normal vector to a plane, so

$$\frac{\partial r}{\partial u} \times \frac{\partial r}{\partial v} = \langle 0, -1, 1 \rangle$$

Therefore, by the Curl Theorem,

$$\oint_{\partial S} \mathbf{F} \cdot \mathbf{T} \, ds = \iint_S \operatorname{curl} \mathbf{F} \cdot \mathbf{N} \, dS$$
$$= \iint_S \langle 1, -1, 1 \rangle \cdot \langle 0, -1, 1 \rangle \, dS$$
$$= \iint_D 2 \, dA$$
$$= 2 \iint_D dA = 2\pi$$

since the region D is the unit circle in the xy plane, which has an area of π.

The Divergence Theorem

The last theorem deals with three dimensional domains, whose boundaries will be a closed surface. It relates the triple integral of the divergence of a vector field to its surface integral on the boundary.

$$\iiint_D \operatorname{div} \mathbf{F} \, dV = \oiint_{\partial D} \mathbf{F} \cdot \mathbf{N} \, dS$$

Example. Compute the surface integral of $\mathbf{F}(x,y,z) = \langle x,y,z \rangle$ over the unit sphere.

Using the Divergence Theorem, we can compute the triple integral of the divergence instead of the surface integral, which makes it much simpler. We don't have to find parametrizations or normal vectors or anything.

$$\oiint_{\partial D} \mathbf{F} \cdot \mathbf{N} \, dS = \iiint_D \text{div } \mathbf{F} \, dV$$
$$= \iiint_D 3 \, dV$$
$$= 3\left(\frac{4\pi}{3}\right) = 4\pi$$

For more complicated problems, we might have to worry about the orientation of the boundary surface. Just as in Green's Theorem, there is a convention for "positively oriented" boundaries. For each boundary surface of the three dimensional region, the unit normal vectors should point *out* of the region.

Exercises

1. Show that if the surface S lies entirely in the xy plane with upward orientation, then the Curl Theorem reduces to Green's Theorem

2. Use Green's Theorem to prove the equality of mixed partials for a two variable real valued function $f(x,y)$ (*Hint: it involves the gradient vector*)

Answers

1. If S is just a flat region in the xy plane, the surface integral becomes a double integral. The unit normal $\mathbf{N}$ is just $\langle 0,0,1 \rangle$, so

$$\iint_S \text{curl } \mathbf{F} \cdot \mathbf{N} \, dS = \iint_S \text{curl } \mathbf{F} \cdot \langle 0,0,1 \rangle \, dA = \iint_S \left(\frac{\partial Q}{\partial x} - \frac{\partial P}{\partial y}\right) dA = \oint_{\partial S} \mathbf{F} \cdot \mathbf{T} \, ds$$

2. Apply Green's Theorem to the gradient vector $\nabla f(x,y)$; this vector field is conservative so the line integral must be 0 and

$$\iint_D \left(\frac{\partial Q}{\partial x} - \frac{\partial P}{\partial y}\right) dA = \iint_D \left(\frac{\partial^2 f}{\partial x \partial y} - \frac{\partial^2 f}{\partial y \partial x}\right) dA = 0$$

Since this must be true for any random domain, the integrand must be zero so

$$\frac{\partial^2 f}{\partial x \partial y} = \frac{\partial^2 f}{\partial y \partial x}$$

Made in the USA
Middletown, DE
17 June 2019